Leslie Weatherhead

Books by the Rev. Dr. Leslie D. Weatherhead, C.B.E., M.A., Ph.D., D.Litt., D.D.

Published by Hodder and Stoughton

PSYCHOLOGY, RELIGION AND HEALING
HIS LIFE AND OURS
HOW CAN I FIND GOD?
PSYCHOLOGY AND LIFE
IT HAPPENED IN PALESTINE
A SHEPHERD REMEMBERS
PERSONALITIES OF THE PASSION
IN QUEST OF A KINGDOM
WHEN THE LAMP FLICKERS
PRESCRIPTION FOR ANXIETY
A PRIVATE HOUSE OF PRAYER
THE RESURRECTION OF CHRIST
KEY NEXT DOOR
WOUNDED SPIRITS
THE CHRISTIAN AGNOSTIC

Published by the Epworth Press

THE TRANSFORMING FRIENDSHIP
JESUS AND OURSELVES
PSYCHOLOGY IN SERVICE OF THE SOUL
THE WILL OF GOD
A PLAIN MAN LOOKS AT THE CROSS
AFTERWORLD OF THE POETS
AFTER DEATH
THE SIGNIFICANCE OF SILENCE
THE RESURRECTION AND THE LIFE
THAT IMMORTAL SEA
OVER HIS OWN SIGNATURE
DAILY READINGS FROM THE WORKS OF LESLIE D. WEATHERHEAD

Published by the Student Christian Movement Press

THE MASTERY OF SEX THROUGH PSYCHOLOGY AND RELIGION
WHY DO MEN SUFFER?
DISCIPLESHIP
THE ETERNAL VOICE

Published by the Denholm House Press

LIFE BEGINS AT DEATH
THE BUSY MAN'S OLD TESTAMENT

Published by the Lutterworth Press

TIME FOR GOD

Leslie Weatherhead

a Personal Portrait

by

KINGSLEY WEATHERHEAD

HODDER AND STOUGHTON
LONDON SYDNEY AUCKLAND TORONTO

For Hansie and Norman French

"To make God real; to change men's lives"

Acknowledgments

I acknowledge the use of quotations from the works of Leslie D. Weatherhead, originally published by The Epworth Press, the Student Christian Movement Press, the Independent Press, Hodder and Stoughton Ltd., and in the *City Temple Tidings*. The poem, "Piano", from *The Complete Poems of D. H. Lawrence* is quoted by permission of Laurence Pollinger Ltd. and the estate of the late Mrs. Frieda Lawrence.

I am grateful to various correspondents for letting me have access to letters from Leslie D. Weatherhead; to A. D. Weatherhead and Frank Cumbers for assistance and advice, which does not relieve me from complete responsibility for errors; to Winifred Weddell and Judy Duke for clerical assistance; to my wife, Ingrid, for assistance that does not come under a single rubric; and to my father, of course.

Eugene, Oregon, U.S.A. A. K. W.
1975

Illustrations

Acknowledgments

1 E. W. Tattersall
2 Walter Stoneman: J. Russell & Sons
3 Fox Photos Ltd.
4 Sport & General Press Agency Ltd.

I

On a warm evening in mid-October 1960, in the large church in London, the preliminaries were over. The organist was playing over the melody of the last hymn to cover the rustle and whistle of stiff Sunday clothes, the clap of hymn books returned to their ledges, and the volley of small staccato coughs, as over two thousand worshippers, sitting down, shuffling and easing themselves, looked up to the brilliant white pulpit where, as the noise subsided and the organ trailed off, a tall, robed figure, white-haired with bowed head, moved slowly up the steps and stood before them. This was the moment they had been waiting for—some indeed had actually stood in the street that evening for three and a half hours before the service started, to be sure of getting a place. There were men and women present who had known this moment over and over again for a quarter of a century—the receding sounds and the silence, and into it the faraway rumble of an underground train or the muffled roar of a bus tearing through the empty Sunday city, and then the kind, urgent, rather high-pitched voice which began to talk—to talk of Christianity, bringing its principles warm and living into the daily practical affairs of men and women: "To make God real; to change men's lives." To some of the sightseers in the crowd that evening, this man was the extraordinary phenomenon who still, in 1960, could fill a big city church to capacity and overflow; to many he was their minister; to some, Dr. Weatherhead; to one or two old soldiers, the padre; to others, Leslie; to a few, just Les. The church was the only Nonconformist church in the City of London, the City Temple.

It was his last sermon to the people who had been his people for twenty-four years. And he preached from a phrase of St. Paul's, "According to my gospel": Paul knew he could reveal only that fraction of the gospel that he had made his own; so, said Leslie Weatherhead, he could likewise preach only that which he could see. That he had done, and now it was time to relinquish the pastorate to another. It was a straightforward appeal to the people to support his successor, but the air was charged with feeling, and the church was unusually quiet. He closed quite simply: "I must say 'Goodbye'," he said. "Thank you for all your love and loyalty, your prayer and your support. I shall never forget you." That evening marked the end of a career of preaching begun half a century earlier in a small village chapel. And, curiously, the text he had taken for his sermon then in 1911, as a young man, he repeated in the prayers in 1960: "Come unto Me all ye that labour and are heavy laden and I will give you rest."

He himself had earned some rest after the London years: preaching in the crowded City Temple Sunday after Sunday, week after week, month in month out; presiding over Fellowship, Samaritan League, Literary Society, and other week-night activities; baptising babies, marrying the young, burying the dead; interviewing people distracted by guilt and fears, some whose parents hadn't loved them, some whose children ran amok; reading letters by the score from men and women in distress; answering the telephone; visiting hospital wards, homes with drawn blinds, prisons; writing books and more books, writing articles for the press; lecturing, broadcasting; holding week-night services in the suburbs, preaching on Sundays in the provinces and arriving back late at King's Cross in unheated trains; opening bazaars and sales of work, appearing at charity teas, autographing books, and shaking endless hands; reading theology at night, preparing sermons, praying. And then on top of it all, during his years as minister of the Temple there had been the war—the killing of young men, the sudden deaths by bombs, the upset and upheaval of his people, the destruction of homes, the flying shrapnel in the streets at night, the blackout, the burning of the church, the years of exile in strange pulpits, and all the other associated tribulations and the palpable miseries. He had earned retirement and rest.

Now the heavy daily demands of running a big preaching church were at an end. And the occasion was perhaps not only a personal corner turned: for surely it will be somewhere in these years that historians of church and religion will mark the point when preaching began to decline as the central feature of the service in the Nonconformist churches. Other good preachers have come, for sure, to the pulpit into which my father used to rise so dramatically, but none with that commanding presence that took so much of the burden of Sunday worship upon itself, upon its own acts and words and movements. In the porch of the City Temple stand two busts, one of Les and one of Joseph Parker, who built the church, who from 1869 to 1902 earned a reputation of being its great sensational orator, and who among his last words commanded a friend, "Keep up the pulpit!" Parker, that massive dominating man, is in marble; Les is in bronze, in a modern expressionistic style he didn't much care for. The two men's preaching styles were also different—as different as the twentieth century from the nineteenth—and Les used to laugh at anecdotes about old Parker and his colossal self-esteem. But they had things in common: one can imagine them, marble and bronze, nodding to each other across the porch and across the divide of the years with some mutual respect and some agreement; and although Les had seen the writing on the wall and believed that in fact the role of preaching would diminish, one thing they would have agreed upon, I think, would have been its central value in the Nonconformist pastorate.

But change has set in: now in all the realms of human activity we have fallen upon times when the leadership of the single great individual has given way to the co-operative work of the group, the committee deliberation, the rap session, and other kinds of combined effort to which the idea of personality is alien. The present incumbent of the City Temple pulpit, Kenneth Slack, himself a preacher of finish and distinction, has spoken of the "enthronement of the sermon" as "distorting and even dangerous . . ." And it is another sign of these times that the Kingsway Hall, which, since Donald Soper came to it in 1936, had been one of the great preaching centres in London, has been sold. And I dare say an earlier milestone in the same movement was set up in 1960 when Leslie Weatherhead left London to its hurried devices and, driving down through the autumn countryside, went into retirement on the south coast of England.

We look back always with some self-delusion to a past that seems more tranquil than our own time. The world into which Leslie Dixon Weatherhead was born in the late evening of October 14th, 1893, in London, certainly seems different from ours. It had not yet felt the impact of the things that have mostly shaped what we think of as modernity: it was as yet untouched by the strange revolutions, intellectual, social and political, that have derived from Freud and Marx; it had not yet known the slaughter on the Somme; and technology, such as it was, was still its servant. It was largely a horse-drawn world, though a few years earlier it had witnessed a giant step forward in the introduction of the petrol engine.

For all the eighty-odd years that have flowed between then and now, however, there is much in the London of those days that is grievously familiar, with merely a change of names here and there. In October 1893 the coal miners were on strike; conflict was imminent in Matabeleland, and there was talk of sending out another British regiment; cholera had been reported on the Continent and was suspected in Greenwich; England was building battleships to keep pace with what were tactfully referred to in the press as "other powers"; a group of unemployed men had marched in protest through the City and addressed a policeman in obscene language.

Saturday, October 14th, had been a dull, damp day in London, unseasonably warm. On the previous day the Conference of the Congregational Union had closed, passing a resolution directing attention to the evils of the Sunday traffic in intoxicating liquors, noting that the working classes, "who chiefly suffer by the Sunday traffic", were "eager" for its discontinuance. Next day, the Reverend Joseph Parker was to preach at the City Temple at eleven and seven o'clock; Queen Victoria was at Balmoral Castle in Scotland; *Charley's Aunt* was playing at the Globe.

Leslie's birthplace was in Harlesden, in the north-west of London and at that time almost in the country. But I don't think the actual locality meant a great deal to him; later he never referred to it: when asked where he was born, he would say simply, "London". We must often have passed fairly close to the street and the actual house, or at least its site, when driving out west on the North Circular Road or A40, but I don't ever remember his remarking it: he never said, "Well, Harlesden! I was born hereabouts", and, a favourite sentence of his that expressed all kinds of emotion, "Isn't that interesting!"

He was only two years old or so when the family moved from London to Leicester, and he had no memories of those early days. The feelings he had for London he developed later, in the thirties, when he became a Londoner, and they rose to some intensity in the forties when, having friends and church members spread all over the area, he saw the city burning. But the feelings were not associated with the fact that he was born there. It was rather the Borders of Scotland, his father's native country, that he felt strongly about; he associated himself with the little town of Moffat, where his grandfather had had a baker's shop in the High Street, over which his father, Andrew Weatherhead, had been born.

It is a town of grey stone fifty miles south-west of Edinburgh. The name Moffat is derived from an old word meaning an opening out; and here in fact three rivers come together from the north and the land opens out into a broad valley. It is peaceful sheep-farming country now, and to celebrate its main industry there is a great bronze ram on a pedestal in the High Street; but it is wild enough, and in earlier days it was the scene of border skirmishing between Englishman and Scot. Here and there about the countryside one may see little fortresses, some in ruins, some still standing, where families with their animals and possessions would take refuge from the English raiders. Then, five miles north of the town, where the sides of the valley become steep and it ends abruptly, the area is known as the Devil's Beef Tub; for here into this blind valley the Scots raiders would drive the stolen English cattle as into a natural paddock. Beyond the Beef Tub, near the source of the River Tweed, is the village of Tweedsmuir, where John Buchan was born, who was to become the writer of boys' books, the Governor-General of Canada, and Lord Tweedsmuir. In Moffat itself, as probably in every town in the Borders for that matter, is a pub where Robert Burns had drunk deep; and here on a window he had scratched a little poem with his diamond. To the north is the Gordon Arms, where Sir Walter Scott met James Hogg, the Ettrick Shepherd, whose poetry arose in the early nineteenth century from the sights and sounds of these bleak rain-washed valleys and whose lines to a skylark

O'er fell and fountain sheen,
O'er moor and mountain green . . .
Emblem of happiness,
Blest is thy dwelling-place—
O to abide in the desert with thee!

provided Cardinal Newman with images and diction for his famous hymn, "Lead, Kindly Light".

Les cherished the local history of the area and also, far more, his own personal memories. When he was a boy his family would take occasional holidays in Moffat. He would sometimes hitch a ride on the box of the coach that went up to St. Mary's Loch, where Robert Burns, drinking in the inn with Sir Walter Scott and hearing that the wine had run out, had cried, "Bring in the loch and we'll finish it!" And then he would fish all day in the Moffat Water Valley and hitch a ride home on the coach in the evening. He used to "guddle" the trout—sliding his hand gently under them until he could flick them out of the water. In later years he used to get excited at the memories of these days. During the nineteen-forties and fifties he again took some holidays in Moffat with whichever members of the family were available, each day taking a packed lunch and climbing the hills and the low mountains—Gallow Hill, Well Hill, Hart Fell, Bodsbeck Law, White Coombe, and Queensberry. It meant much to him to get to the top of one of these. Sometimes he would leave a visiting card in a bottle in the cairn at the peak; and once, on a foggy winter evening in London, a visitor to the church told him he had found his card on White Coombe, and momentarily the whole sunlit range flashed before his inward eye. Once Adamson, the town's milkman and taxi-driver, made what seemed to me a fine epithet for him in that curt style of the Scot that says much in little: "Aye," he said, "he kens the tops." Well, he did. He knew them and loved them and was in awe of them. Nothing in nature moved him like a mountain range, and in his writing and preaching, indeed in his thinking, the mountain regularly served as an analogy for spiritual heights, or divine attributes, or whatever he held in unlimited reverence and admiration. His love of mountains and those qualities for which they stood was sometimes exclusive and separated him from enjoyment of the rich dull and daily levels of human life. Rilke spoke of tables and chairs and crockery as elements in which humanity is stored: Les wouldn't have seen the point of that at all. Much of his life he lived figuratively among the mountain tops; and when in age things wound down, when he was widowed and in poor health and confined to small circumscribed routines, he couldn't generate any enthusiasm at all for the ordinary texture of his days, and he wanted to die: the mountains were no longer available, and nothing else would do. So in the passing comment of the milkman, "He kens the tops", there was a symbolic truth that he did not realise.

"I should like to think," Les writes, "that my ancestors were shepherds on the hills I love in the Border country of Scotland." Perhaps way back his people had been shepherds, or cattle thieves—another imaginary ancestry he found attractive; but the record shows that since 1700, at least, Weatheretts, Watherheads, and Weatherheads had been in turn baker, fewar (tax collector), weaver, and baker again, though of course some of them may have done a bit of cattle rustling in their spare time without even registering as amateurs.

Andrew Weatherhead, the son of the baker in Moffat, had been educated at the Moffat Academy, one of the local Scottish secondary schools which have the reputation of being much superior to their English equivalents. His two brothers were also there: John, who became manager of Barclay's Bank in Birmingham, and James, who became a minister and later the Moderator of the United Free Church of Scotland. Andrew had come to London to work in a hosiery warehouse. Les spoke of him as a man of inflexible honesty and integrity, and he remembered one small application of his strong principles: rather than pay a tax of which he disapproved, when it came due his father suffered the police to enter his house to take away a small article, a clock for instance, or a lamp, which he would then later redeem with the money. When he became manager of T. H. Downing, a wholesale hosiery concern in Leicester, if with great reluctance he fired a man, he would visit his house afterwards to make sure the man had enough to eat. He must have been a shy man whose goodness awkwardly and embarrassingly appeared in acts of covert kindness. He would have wanted no more theatrical a tribute than what was repeatedly paid him in the speeches at his retirement dinner in 1928 in the designation, "gentleman". But he was too reticent a man for a boy to confide in. There was, no doubt, the distance between child and parent, familiar in the annals of Victorian life, which made it difficult for them to know one another. But he helped his son make an electric bell once, and helped him fix up the attic as a darkroom for his photography; in his own stiff and formal way he showed he cared for him.

In London, Andrew Weatherhead attended Wesley's Chapel on Sundays. There he met Elizabeth Dixon, a school teacher, who became his wife. It looked, many years later, as if this pulpit were to be offered to Leslie, then a young and startling new voice in Leeds, and it would have come to him with great delight—"The church," he used to remark, "where my parents touched hands in the back pews." But his appointment was blocked by a fellow minister in the Methodist Conference, who was old, reactionary, and powerful.

The Andrew Weatherheads had three children, Muriel, born in 1888, Alice, in 1890, and Leslie, in 1893, who of the three alone survived middle age. The mother, Elizabeth, was a Christian of terrible and serious aspect. In the old sepia print of her we see a handsome woman of fine figure with rich hair piled up; but the face is sharp-featured and stern, the mouth is firm, and there is not much joy in the eyes behind the pince-nez. She was, Les used to say, a "dominating woman". The words belong in quotation because for Les they designated the kind of human being that he most disliked on earth. She was austere with herself and with others: in the room where she lay dying she was concerned with all the flowers she had been given: "People think too much of me," she complained. She seems to have been a great forbidder; no one in the family could enjoy what she disapproved of, and what she disapproved of covered a wide spectrum of recreations—dancing, smoking, drinking, gambling, theatre, cinema, and a number of other pastimes for weekdays—the list of Sunday prohibitions being, of course, longer.

But if she seems to have clamped down on many of the activities that gladden the average secular human heart, she was not, in these disciplines and denials, by any means egregious among the Protestant middle class of England in those years. There must have been thousands in her social situation, for instance, who shrank in dull distaste from the theatre, which had only just begun its slow climb to respectability when, two years after Leslie's birth, Henry Irving became the first actor to be knighted. Nor is it surprising that she should abhor drink. In fact, she was an active temperance worker; but again, in her social situation probably the only non-teetotallers she would have observed would have been the drunks staggering past the area railings as they left the local pub at closing time on Saturdays. One night, walking with his father, Les saw a man return to his own doorstep, fumble for his keys, and then in impatience raise his cane and smash in the glass of his own front door. His father squeezed Les's arm and whispered, "*That* is what drink does!"

For a long time Les's attitude to drink was fashioned by his parents'; years later, when we used to drive through Doncaster, a rich odour like cooking parsnips would come wafting into the car and Les would sniff and mutter about the perniciousness of the brewing trade and all its works. "One glass of beer," he would quote, "lowers your speed of reaction by 20 per cent." He was quite fond of statistics. The attitude of his friends was most often the same as his. Quite recently, for example, Donald Soper said, "I'd close every pub in Britain." Les did not retain intact the chilly attitudes he had inherited, though they monitored his practices throughout life. They were slightly modified in the 1940s when Lord Horder, his doctor, ordered him a glass of port after lunch on Sundays. We children thought he was the right kind of a lord—getting the old man mellowed up on the sabbath! My mother was always strictly teetotal, and when she was ordered burgundy by *her* doctor, she mixed it with the water in which beetroot had been cooked (whether for reasons of pharmacology or masochism, I don't know). We all thought that was a pretty awful thing to do, but at least the burgundy wasn't a very good year.

Later in his life Leslie realised that his mother had not wanted him. When he arrived he was a difficult baby, and he became a delicate and sickly child. As a boy he feared his mother, and yet he earnestly desired her love. She sang evangelical hymns as she swept the stairs, and she beat him whenever he broke the rules. He never forgot the repeated beatings. One day at her housework she heard the crash as the boy catapulted a marble through the glass of the little conservatory they had against the back door. The sweeping stopped and the strains of "Shall we gather at the river?" abruptly died. The mother picked up a cane and beat the boy all the way up to his room. There she stripped him and again beat him violently. When she had finished, he lay in bed and cried, watching her in the dressing-table mirror as she pinned up her hair which had fallen during the assault. He hated her, and, as the highest pitch of terror is reached not by what others do to us but by the prospect of what we may do to others, he trembled with the fear that his own hatred and desire for revenge had induced in him. But above all, he wanted her to love him. Sometimes, after he had been punished, she would withhold forgiveness; and the boy would sob for hours until she reassured him of her love. Often in these memories, which lived with him for ever, the mother seems to us to have been almost inhuman. And yet, later, it is quite obvious that she thought the world of him, in the old fading letters that she had written to him during the war, which begin, "My own dear Laddie." He, for his part, though he retained all the dreadful memories, remembered also that when he was a child he had thought she was a marvellous person. And when she lay dying in 1921 and he was unable to be at the bed-side, he writes in his diary, "My heart goes out to her; I long to be with her; we were such pals." He repeats the sentiment two or three times throughout the following years. In 1927 a prayer in his diary opens, "O God, my mother's God and though I am unworthy my God still . . ." And then in the next year when he came to London to preach for the first time as a guest at the City Temple, he had an extraordinary experience which involved a memory of his mother: apprehensive about the ordeal of Sunday morning, he awoke the night before to an overwhelming sense of her presence and was filled with joy. "I shall never forget that sense of exaltation and well-being," he writes. "All fear was taken away. I cannot use a word like transfiguration. I can only speak of a heart that sang and a spirit that was in ecstasy." With such associations of his mother's memory it is hard indeed to untangle the mixed skein of healed injuries and raw recollections to come to a clear conclusion about his real feelings. What do we ever know, for that matter, about the deep make-up of a human heart?

Long before the boy was old enough to operate a catapult, the family had moved from London to Leicester, where Andrew Weatherhead had been engaged by the wholesale hosiery firm of which he was to become managing director. Leicester was a centre for woollen goods as it had been for six or seven centuries, and hosiery was its staple trade. The Weatherheads moved to a modest street befitting their station in life. We can picture the house with the conservatory against the back door, the black pointed iron railings in front; the stone steps with their sand-stoned edges that led up to the darkly painted front door; the polished brass of door-handle, knocker, bell-push and letter-box; the prim front-room windows with the blinds half-drawn and the lace curtains opened only far enough to admit what little light was necessary for the culture of the aspidistra that stood in its brass bowl on the table inside.

The front room would be reserved for visitors: it would be kept dusted, and the antimacassars on the overstuffed furniture would be kept straight and the cushions plumped for the occasions when Leslie's mother poured tea for her friends and the co-workers of the Wesleyan church that the family attended. For Elizabeth Weatherhead lent much of her strength and vitality to such departments of the church as the Wesley Guild, the Young Women's Class, Foreign Missions, the local branch of the Women's Temperance Association, and the Sunday School, where she taught for twenty years. In a sense beyond that which usually applies in obituaries, she led a life devoted to the church.

Needless to say, her devotion found expression also in varieties of Christian exercise in the home. Family prayers, which were the custom among most upper and middle class people in the late nineteenth century, were routine in the Weatherhead household. Every morning after Andrew had gone off to work—after he had "escaped", as Leslie later expressed it—Elizabeth would read aloud to the rest of the family the verses of the Bible prescribed for the day by the International Bible Reading Association, and then she would read the commentary that the Association had supplied. Then the four of them would kneel while she offered extempore prayers, and then they would say the Lord's Prayer. There were special prayers for special occasions: on mornings when he had exams at school, Les remembers, his mother would request the Lord to "give him that measure of success which Thou seest best for us all". "If that prayer was answered," he comments, "then what was 'best for us all' was a very modest requirement."

On Sundays the morning routine was similar, except that it was the dining-room floor and not the cold flagstones of the kitchen that they knelt down on, and Andrew did not escape. But in addition there was Sunday School twice, chapel twice, and prayers again in the evening. Between tea and evening chapel, their father taught them the Shorter Catechism; "What is man's chief end?" it required them to know, and "What is effectual calling?" Evening chapel was followed by the prayer meeting, at which a number of the worshippers would remain behind after the service for prayers from the floor of the chapel. Elizabeth would whisper, "You pray, Andrew." But he rarely did; she, frequently. When they returned home there were hymns round the piano before the last family prayers, and then at last the lonely individual supplications when the boy's knees bent once more, this time on his bedroom floor.

Once when he was old, lying back as he used in his favourite study chair, for some reason I cannot now remember, I read him the poem by D. H. Lawrence called "Piano":

Softly, in the dusk, a woman is singing to me;
Taking me back down the vista of years, till I see
A child sitting under the piano, in the boom of the tingling strings
And pressing the small, poised feet of a mother who smiles as she sings.

In spite of myself, the insidious mastery of song
Betrays me back, till the heart of me weeps to belong
To the old Sunday evenings at home, with winter outside
And hymns in the cosy parlour, the tinkling piano our guide.

So now it is vain for the singer to burst into clamour
With the great black piano appassionato. The glamour
Of childish days is upon me, my manhood is cast
Down in the flood of remembrance, I weep like a child for the past.

It is a good poem by any standards. But the electric effect it had on Les was quite surprising. As it unfolded he sat upright and wide-eyed in his chair. "That's me! That's exactly me," he said. "'The old Sunday evenings at home'."

There must have been thousands of people like my grandparents for whom Sunday meant prohibitions and discipline. It seems a different world from ours in which children read the comics in the Sunday papers while watching simultaneously some clown on television chanting hosannas to a brand of motor oil. In our parents' lives there were things that we don't have: they had firm values that measured men's worth by serious, sober standards; they knew vice from virtue, and they had no special terminology for evil. But there was much deprivation of the senses: the beauty of the world, some of the simple pleasures of ear and eye, the beauty of person and the gracefulness of act, none of these were to be countenanced or appreciated before they had been scrutinised for their moral implications. If they believed that beauty was truth, they kept it a pretty dark secret.

For Les, Sunday was probably a day on which he was less lonely than on others, when he was of pre-school age at any rate. For much of the time in those years he was lonely. Sometimes his sisters, both older than he, took pity on him and would permit him to play with them: he could be the doll-doctor and succour the dolls in their ailments. But dolls, alas, do not suffer from lingering and chronic diseases; and, his part quickly played out, Les would be dropped from the game. Perhaps it was on such occasions, excluded from his sisters' play, that he first entered upon his own golden childish imaginings, climbing on to a kitchen chair, raising his arms over the back and, as if flicked by a glimpse of the future, preaching to his massed congregations. His mother took much satisfaction from this game; she had dedicated him to God before he was born and had nourished the hope that he would become a minister.

One image of uncertain date, recalled from early childhood, concerned these things: he had been sent to the little local shop to buy greengroceries and stood before the window, holding the basket. And there, staring at his reflection superimposed upon the cabbages and carrots and vegetable marrows, he said aloud, "I could be great for Christ." Another incident from childhood that foreshadows later things is dated: on January 3rd, 1903, he determined to serve Christ for the rest of his life, and he noted the fact in red ink in a new diary he had been given for Christmas.

By that time, however, he was aged nine. Before then he had been sent to a day school, an institution which came to live in his memory as a horror. The first day, as his mother escorted him to the infant department and they were ushered into the presence of the headmistress, this executive, a white-faced, unsmiling woman, was engaged in beating four children. Les remembered the sound of the cane which whistled as it came down and made a sickening thud as it struck the soft hands of the boys and girls, who shrieked with the pain. This, at the age of five, was his introduction to formal education. He was soon to learn that this scene was no unusual ritual at the school: if a child was late he was automatically beaten, whether he had an excuse or not; if he failed to understand his lesson he was beaten, and it remained unexplained; if he made errors in his homework, he was beaten. Leslie's recollections make the school seem like one of those awful places run by scoundrels in Dickens. It must at least have been a singularly cruel establishment. From all the eight years he spent there, he could dredge up only one good memory: returning one September after the summer holidays he saw a notice on the gate announcing that, owing to an outbreak of smallpox, the school would not reopen for three weeks. He didn't know what smallpox was, but thereafter towards the end of each holiday he prayed to God that there would be another epidemic. But God's purposes are inscrutable, and never again did Divine benevolence manifest itself in this way! Leslie hated the school so much that he used to go about with wet feet, even pour cold water over his feet when going to bed at night, in order to induce colds and bronchitis, and he became a fairly regular victim. Once, being ill with bronchitis, he missed all the instruction that had been offered in long-division. But then, determined to master these mysteries, he had his sister assign him problems at home, and he sweated them out all on his own. When he returned to school, however, and took his answers to the teacher, she said, "Well, anyone can get his sister to do his sums for him and then just copy them out." With which she tore the paper in half with her fat white hands, and picked up her cane for the academic improvement of the next child in line.

The injustice of this incident rankled. But it was the beating that had the most lasting effect on Les: as a boy it instilled in him continual terror and humiliation; and then as a man for the rest of his life, whenever he heard or read of the corporal punishment of boys and girls, it aroused in him an irrational pitch of anger. In the 1930s, when he lived in Leeds and was writing an article each week for the *Leeds Mercury*, he carried on a one-man campaign against beating in the schools. He raked up instances of sadism and other miscellaneous perversions and cruelties. And in exposing some of this squalid stuff in his articles he did some good, in spite of the abuse in the press, the protest meeting that the enraged school teachers held and their vote of censure upon him, and in spite of the inconvenience of coming out of a meeting late at night somewhere on the edge of town to find that the air had been let out of all his tyres.

When he was thirteen Les was entered in Alderman Newton's Secondary School, where some years later the fragments of two cultures were to be introduced to a boy called C. P. Snow. Leslie stayed there for five years. His memories of this school are not particularly happy either. Twenty-five years later, distributing the prizes at the school, he reminded Mr. Newton, one of the masters, how he had given him "six" for talking in an examination. His academic achievement seems to have been moderate. But there was one glorious moment of surprise and triumph in his last year when he learned that he had won the popularity prize, which was awarded according to votes cast by all the boys in the school. He was away from the school sick when he heard the news, and in disbelief he ran round to the house of a schoolfellow to see if it was true. No one who knew Les as a man and had felt the extraordinary charm and magnetism he exerted would have been surprised at the election. But he was only seventeen at the time, and this was the first formal indication of the warmth of popular affection that was to attend and foster so many of his undertakings for the rest of his life.

Leslie's memories of his schooldays are no doubt unconsciously edited and adjusted, as all our memories are, it being part of our human self-deceit that, without knowing we do so, for certain images we hold the door and for others, with credentials no less valid, we close it fast. Surely there must in those years have been some occasions of pleasure and outcroppings of good luck—in addition to the smallpox epidemic and the popularity prize! There must have been some good days when x failed to elude him and his algebra came out right, some days perhaps when, with fortune smiling, he was unobserved pricking the neck of the boy in front with his pen. The catapult missile must once in a while have been on target. It was in these years, presumably, that he developed his zest for association football, "footer", as he always called it. One can imagine the schoolboy scene—the vivid jerseys in the winter sunlight, the slashes of mud on shorts and socks, the raw knees, and the clamour of boys' voices echoing back across half Leicestershire into the rising mists of the afternoon; and one can easily see how that scene translates into another one, forty years later, of a Saturday afternoon, a distinguished gentleman of the clergy, no longer young, wrapped up against the cold in a bulky dark overcoat, dark knitted scarf and black homburg, at the edge of the football field in the little park in the Borough of Finchley, N.3, charging up the touchline, his walking-stick held like a bayonet, roaring to the red-headed wing forward, at the top of his voice, which was better suited to holier offices, "Go it, Ginger, go it! On your own, Ginger! Shoot! SHOOT! GINGER!" reliving boyhood in his fifties, momentarily defeating middle age and its harsh circumscriptions.

In those years also there were experiences that held more meaning for the future than football did. All his life Les loved to read in bed. When he was a boy, G. A. Henty, R. M. Ballantyne, Conan Doyle, and Rider Haggard were the writers who provided adventure stories for boys. And surely Peterkin and Jack, Allan Quatermain and Sir Henry Curtis, with all their impossible feats of courage, inhabited the mind of the boy in the cold back bedroom in the English midlands, who many years later, when he sailed for India as a missionary, was to pack in his luggage, along with the celluloid clerical collars, the thumbed Bible, and the notebook of original prayers written out in careful longhand, a brandy flask and a revolver.

The Emperor Julian saw Christianity as appealing to contemptible men, but Les was to describe later how, when he was a boy, Christianity had become identified in his mind with adventure, and Christ with the heroic leader: "What a Captain He must be," he wrote, "if Livingstone would go through Central Africa for Him, if Paton would go to the New Hebrides for Him, if Calvert would go to the cannibals of Fiji for Him, if Gilmour would go to Mongolia for Him, and if Carey would go to India for Him! What a Captain to sail under!" He could remember also how in the cold bedroom where he fed his dreams his father with a grave Presbyterian face had taken away from him one Sunday night *The Pirates of Acre*.

But then again, it was not only ideas and images of adventure that the boy was gleaning from books. In the First World War, adventure was to come to him anyway—adventure that would do justice to Rider Haggard or *Coral Island*. What came to him in the novels along with the heroic deeds, to be of long significance in his life thereafter, was the power of words. We cannot doubt that some of the gaudy prose sent shivers down his spine as was intended, and that he learned quite early, by intuition, what words could do—the magic they could exert upon him, and then later, in turn, the magic that he could exert through them. It is inconceivable that, as he read these stories or, indeed, as he daily heard and heard again the superb prose of the Bible, the shape and sound of words and something of their mystery with which he was to be concerned for the rest of his life had not begun already to engage him. For always he loved words. He loved the sounds of them and their rhythms, the effects they brought when they were perfectly placed; the nuances and scruples of meaning they could be made to convey. Once in a sermon he took a little time out to discuss the respective merits in different contexts of "window" and "casement". On another occasion he taught his flock the etymologies of "tawdry" and "posh". Mostly he liked words that denoted peace of one kind or another. He used to like to try out words, tasting them on his palate. When we boys grew old enough, my brother and I would tease him a bit about this (and, it may be added, about a thousand other idiosyncrasies): "I think 'quietude' is a beautiful word," he might say. "'Qui-et-ude'. And 'serene'. 'Serene' is a lovely word." We would say, "How about 'blunderbus'? It's got an almost perfect rhythm. And, Les, you know what really sounds good? 'Celery root'!" Or my brother, who wanted more than anything in the world to own a motor bike, would assert that to him the sweetest sounds in the language were "Triumph Speed Twin"! Quite often the sounds and rhythms and the general aura of words meant more to Les than their meanings. "I've been spread-eagled in a nullah," he announced once, by way of accounting for his delayed arrival at a picnic rendezvous in the Moffat hills. "You've been what?" "I've been spread-eagled in a nullah!" Enquiry revealed that he had got into difficulties climbing around in a ravine. Not knowing the Hindi word, someone asked, "Which ullah was it you were in?" "Nullah," Les replied, "n-u-l-l-a-h", introducing us all to this exotic import, for which English has a perfectly adequate supply of domestic synonyms. Later it became quite the thing among the merciless family, if he were late coming down for lunch in the London house, to ask rather pointedly if he'd been stuck again in a nullah. He used to say he had started to write a novel but he had only got as far as the first sentence, which was, "The great limousine swung down the drive"; or maybe it was "purred". I don't think he knew why it was swinging or purring down a drive or what else was going to happen later in the novel: there was the romantic atmosphere surrounding the great car, but after that the rhythms, the sounds, and the diction were everything.

Of course, in his long career as a man of words he was concerned with a great deal more about them than the mere epicurean delight they could offer. It was the self-imposed task of his fifty years' ministry to render the facts of the Christian story in as simple a language as could convey them, and his eminent success in this must concern us at some length in the course of these pages. But we may note here, as we consider the influence that the magic of words had upon the growing boy, that as often as the rationalist in him protested the meaninglessness of institutionalised prayers or the unsolvable mysteries in certain phrases in the litany, learned and repeated by rote, just as often, in sermon, article, or book, the poet in him would choose words of vague imprecision, not to be interpreted so much as felt. I think Les used more magic in his language than he permitted Archbishop Cranmer; and although he proclaimed with some verve that there were parts of the Book of Common Prayer that didn't make sense and could not be stomached, the mysteries in Alfred Lord Tennyson he relished uncritically.

But to our tale. Somewhere towards the end of his time at Alderman Newton's School, Leslie began to develop the idea of becoming a minister. Then he notes in his diary that on Sunday, September 3rd, 1911, he delivered his first sermon in a little village chapel at Houghton-on-the-Hill, six miles out of Leicester. To get there he had to drive a pony and trap for the first time, and he was terrified it would founder. We can well picture the setting for the service—the little chapel with wooden pews and pulpit, a text in Gothic letters on the end wall over the communion table—"Let There Be Light", maybe, or "In the Beginning Was the Word". There was probably only a small gathering—a few farm-workers, men and women, a gardener or two, a maidservant, and perhaps the local school teacher at the pump organ, the kind she would have to bring wheezing into life by working the pedals; outside, the singing of a blackbird and the crying of sheep across the valley, rivalling the clear voice of the pale, black-haired boy not yet eighteen, preaching on the text, strange enough for his youth, "Come unto me all ye that labour and are heavy laden and I will give you rest", so nervous that as he talked he picked at the cord sewn around the cushion on which the Bible rested, until inch by inch he had it all off.

A little later, he notes that he and his father had gone to hear an address by a missionary returned from India. Les had himself earlier entertained the notion of going as a missionary; but, writing some years afterwards in a letter to the local press, he seems to have identified this as the occasion when the call of adventure, "to a boy who revelled in Henty's books", had combined with the call of God to summon him to the mission field. After the meeting he reports that he and his father had missed the train, had walked a long way and had had a "ripping talk". It was a rare enough occasion, and Les took advantage of it to tell his reserved father of his ambitions to become a medical missionary in India. His father did not discourage him, but told him he would have to choose between medicine and the ministry; he could not afford the training for both.

Les then committed himself to the Indian mission field. He began to prepare himself and sat the preliminary examinations in theology and Bible and preached some trial sermons. Early in September 1913, after two terms at Cliff College in Derbyshire, he was installed in Richmond Wesleyan Theological College as a candidate for the ministry. It is a proud-looking building standing in its own grounds over the river Thames, some ten miles upstream from London; and here, one may say, life began to open up for him. He admired the men who taught at the college; he enjoyed a lot of the subjects—theology, church history, homiletics, Hebrew, comparative religion, New Testament Greek, logic, and ethics. He had his own room, his "den" as it was called, and he delighted in the comfort and the seclusion it offered, with the curtains drawn against the night, a shaded lamp, and a coal fire. At the same time he was thoroughly outgoing and gregarious and quickly made many friends among his fellow students. Years later he recalled as one of the chief sources of the happiness of those years the fact that there was not one man in the college with whom he did not feel at home. And his fellows, for their part, showed their estimate of him by electing him, in his first year, secretary of the year; in his second, chairman of the year; and in his third, chairman of the college. At this last election, he was greatly elated: in his diary for June 1915, he calls it "the greatest honour of all my life", and he feels unworthy to follow his predecessors. He hopes he will do well and maintain the high tone of the college.

There were other joys during these years, and there were moments of memorable ecstasies and defeats. Fifteen miles from Leicester, in the Charnwood Forest, was Onebarrow Lodge Farm—some good acres under grain, hay, and pasturage, with a lake and stream and trees and rocks. The Weatherhead family spent a summer vacation at the farm when Leslie was in his teens, and thereafter for a number of years he would often return. There was no episode from this time of his life that in later years he would rather return to, to talk or dream about, than the sights and sounds of Onebarrow Lodge. Later, when he saw pine trees in India, he remembered the pines of Onebarrow; he heard doves in Mesopotamia and he thought of the doves of Onebarrow. But in addition to sights and sounds there were the glorious activities: the haymaking, the loading up of the horse-drawn truck at harvest time, driving the pony and trap, going after the rabbits and pigeons with the twelve-bore, watching the birds and wild animals, chatting up the farmer's daughters or the stable-boy as he polished the buckles of the great shire harnesses, eating enormous meals, and at night taking the candles up the old oak stairway—for any boy, they would have been great days; for a town-bred boy, very heaven.

Leslie says that it was at Onebarrow that he first came under the strong sway of nature, which did not cease to absorb him for the rest of his life. Never have I known anyone in whom the sights and sounds of nature, especially its peaceful ones, could induce such unalloyed pleasure. All through his diaries, his sermons, and his books he presents scenes from nature, English nature, most often scenes at evening of calm landscapes and still trees, but wild scenes also, of sea or cliff, mountains or gorges, where the hand of man had left no imprint. When he was in India, his memories of home were regularly of English nature, and he cut out the plates from magazines that portrayed simple country scenes of England and framed them for the walls of his foreign bungalow. Others, no doubt, have done likewise in nostalgia. But Leslie's strong love was unusual: it was akin, I think, boy and man, to that of the poet William Wordsworth, whose lines will serve admirably to describe it:

> The sounding cataract
> Haunted me like a passion: the tall rock,
> The mountain, and the deep and gloomy wood,
> Their colours and their forms, were then to me
> An appetite . . .

Like Wordsworth again, to the sights and sounds of nature Les owed "another gift",

> Of aspect more sublime; that blessèd mood,
> In which the burthen of the mystery,
> In which the heavy and the weary weight
> Of all this unintelligible world,
> Is lightened:—that serene and blessèd mood,
> In which the affections gently lead us on,—
> Until, the breath of this corporeal frame
> And even the motion of our human blood
> Almost suspended, we are laid asleep
> In body, and become a living soul:
> While with an eye made quiet by the power
> Of harmony, and the deep power of joy,
> We see into the life of things.

You could say all that—yes, all of it, I think—about Les.

There were granite rocks at Onebarrow surmounted by the pines and jutting out over the lake. Leslie called this his temple, and he meditated there in the evenings, looking far away over the rolling wheatfields into the darkening sky, and dreaming his dreams. He came here too with his cousin Louie Weatherhead, with whom he was in love. There had been other girls before her—a Nellie, a Gladys, and an Elsie—whom he had liked well enough and had so informed. This was the real thing, however, and he wanted to marry. But they were close of kin, and they had to recognise that they were barred from marriage. It was impossible. But it was great, it was tremendous to be linked arm in arm with her, walking through the fields at Onebarrow in the evening when the rooks were flopping back to their nests in the tops of the trees. When they parted, they exchanged letters: his speak of heartache and despair; hers are exhortations to be sensible. They determined to be good friends, just very good friends. But for his part, the relationship did not soon descend to one of Christmas greetings and holiday postcards. And years later in India he records in his diary that letters from Louie still stir memories and desires. And in 1918 when she reported that she had become engaged to be married, he felt a shock of pain and wondered whether he should ever love another girl as well. In his distress he reached for his Tennyson and found comfort in that sentiment that evaluates defeat in love higher than non-participation.

When he became chairman of the college and made his private pledges to serve to the best of his ability, Leslie did not know how little opportunity he was going to have. In September 1915 his diary records that he is back at Richmond, moved into the chairman's den, again relishing the privacy of the room, the companionship of the men, and the discussions they had together over their pipes late into the night. But in October he learns that, according to the characteristic illogic of wartime measures, Richmond is to be taken over by the people from Westminster Training College, whose premises had been commandeered in turn by the Australian Government for some purpose related to the war. The first- and second-year men at Richmond were moved to a college in Manchester, but the third-year men, Les among them, were given preaching appointments. He was sent to Farnham in Surrey, a town in an urban area forty miles south-west of London on the road to the New Forest and Winchester.

Here began his ministry: aged twenty-two, with no degree and unordained, he was in complete charge of a church in the town and also of three churches in the country. He lived in the town with a family with daughters, and in spare time he exercised their airedale dog in the lanes and birch woods of the county. There was not a lot of spare time, however. The young novice had to make sermons for Sundays; he had a Society Class of forty girls to lead, week-day services in the villages, prayer meetings, the Wesley Guild, and families to visit on his spluttering second-hand motor bike. Some of the miscellaneous activities were dutifully recorded in the press: "The Rev. L. Weatherhead having led in prayer," we read, "said perhaps there never had been a more difficult time for holding a sale of work," though "the workers had done splendidly." In another clipping we read that the Farnham Wesley Guild offered entertainment for wounded soldiers in the schoolroom, over which Leslie presided and also accompanied Miss Fostekew and Mrs. Menday as they sang a duet. At a tea for raising money for renovations, he "moved a hearty vote of thanks to the choir . . ."

But the yellowing press clippings of those days begin to outline also another story: one reads, "Although so young—being fresh from Richmond College, he is an acccomplished preacher, and draws large congregations . . ." At the Home Mission Meeting he had "a crowded congregation"; at the Sunday School Anniversary service he "based an excellent address on the words, 'keep smiling'"; at a local funeral he "spoke impressively"; we read that he gave a "splendid sermon" here and a "fine address" there.

Without question he was already making his mark, a happy man among congenial people in a job to which he was obviously eminently suited. So it must have appeared. But in those days there was a deep undertow of unhappiness for everybody in England, for men of military age and men of conscience, at any rate. Les knew as well as anyone that the times were difficult, not just for holding a sale of work. We have seen in our own day only too closely the distress of young men whose principles are outraged by the demands of the State, and we have heard their voices raised. In 1914 in England there were arguments among the young men: at Richmond they had recognised the particular awkwardness of their position as spiritual leaders, and these were matters they discussed in smoke-filled dens night after night into the small hours. But the First World War was different from Viet Nam: it was the war to end war—one could believe this then; it had songs and righteousness going for it; and the principles of most men of principle prompted them to join the army and fight the evil foe rather than take a pacifist stand. Leslie himself seems to have been strongly conscious of the duty of serving in the army. The record in the diaries in 1915 is not quite clear on these matters, but the confusion is due probably to the hyperactivity of a conscience, contemplating perhaps the awful spectacle of British losses in the recent battles over the Ypres salient. Before going down to Farnham he had reported to the recruiting office in Leicester but had been rejected on medical grounds. Then he applied for a chaplaincy but was rejected and told he was already doing chaplain's work in Farnham. In his diary for November 1915, recording his utter abashment before the extraordinary phenomena of the day, he lists among these "men going the enviable khaki way", as if this had been denied him, as it seems it had by the medical officer. Three weeks later, however, he is concerned with the matter of Duty—with a capital D—arguing with himself, about it and about trying to distinguish temptation from principle and his own will from God's. The problem obviously nagged him, and probably not least because he suspected that his immediate superior, a minister in Aldershot, despised him for not being in the services. It was all solved in the end, and, as so often with such things, it was solved in a way he could hardly have anticipated. Meanwhile, his commitment to India was to be implemented: in the summer of 1916, he learned that he was to be appointed to Madras and was to sail in the autumn. He notes in his diary, "Glory be to God. This is the realisation of the ideal of all my life."

2

The *Derbyshire* was due to sail on October 12th, 1916, and Les's parents and his younger sister, Alice, journeyed from Leicester to Liverpool to see him off. The boat was delayed. The family waited in the town and returned again the next day to the dock, but still the boat did not sail. It left finally on the fourteenth; but by then, having said their painful goodbyes twice over, the family had gone home, and Les had seen his mother for the last time. It was his twenty-third birthday. When the ship pulled away from the harbour there was no one to whom he could wave. He must have felt the loneliness to which he was always prone. Often in later years he would recall the anecdote about Rupert Brooke who, on leaving England with no one to see him off, dashed down the gangway just before the ship weighed anchor, found an urchin on the pier, gave him sixpence and a clean handkerchief and bade him wave; then long afterwards, when the pier had been deserted and was dissolving into the sea haze, Brooke could still see his small diminishing stalwart conscientiously waving. Probably, on his leaving, Les did not linger long at the rail, as the harbour with its derricks and smokestacks and the blackened buildings of Liverpool receded. The world lay all before him, bearing the great adventure for which his life thus far had been preparation. Before him more immediately lay the seven-thousand-mile voyage to Colombo, Ceylon.

They had a storm in the Irish Sea to start them off, and then there were the amazing views of the Rock of Gibraltar and the coast of Spain; there was a visit to Marseilles. There was Port Said, and then as they sailed into the moonlit Suez Canal Les felt above the muffled thud of the engines the encroaching silence and the mystery of the deserts. There was Aden, the dusty coaling station, and after that the long hot haul across the Indian Ocean.

It was an interesting trip for a young man who had never been out of Britain, and there were German submarines about to add to the excitement only a few days before the *Derbyshire* sailed, world indignation had been aroused at the sinking of the Dutch ship *Blommersdijk*, an unarmed neutral vessel carrying wheat. For all the potential adventure, however, Les found that time weighed heavily on his hands: he was eager to get to India and get to work. There were the usual activities on board, designed

Les's mother.

College student at Richmond, 1913.

Richmond College missionary platform, 1914.

Missionary in Madras, 1917. Aged 24.

Chaplain, 1918.

Evelyn Triggs, missionary in Madras.

to kill the hours: deck tennis, shuffleboard, dancing, bridge. And Les entered into these diversions with some enthusiasm: he even played Sibelius's "Valse Triste" on the violin, accompanied by the ship's orchestra. The fun and games, trivial in themselves, helped to relieve the boredom inasmuch as they brought him into contact with the other passengers. A fellow missionary aboard criticised him and reported him to the Missionary Society for his worldly behaviour: it was probably a little French girl whom Les danced with who particularly aroused the indignation of the colleague, and Les was a little penitent. But only a little: cloistered virtue was not for him, and furthermore he wanted to get to know his fellow travellers in order to persuade them into a Christian commitment—the French girl being perhaps an exception, her soul not his first concern. For better or worse, she disembarked at Marseilles.

On board he was beginning to meet people whom he found most agreeable, and it worried him very much to discover that they were not, in his words, "in touch with God". In his diary of the voyage, he registers his concern for their spiritual condition: he is pained to report that on Sunday morning after the Church of England service, poorly attended as it was, the cards and skittles started up in full swing. Among his aquaintance is named a Mr. Milton who seems to be okay spiritually; but Mrs. Turton, Mr. Labelle, and Miss Wenyon cause him concern. He reports that he has had "deep talks" with each of them; but he found that though they were charming people, and had much good in them, they did not seriously accept Christianity; they were not, as he says, "anchored to reality".

Men with consuming ideals for export are liable to endure the snub, or the retort, or cool indifference when they make their pitch. And for all the wit and charm of this young man, getting the "deep talks" under way must have occasioned some embarrassment among the nice people who, he found, "appealed to him tremendously"—a tea-planter and his women, perhaps; a civil servant; a military family. We can imagine the earnest young man trying to force the moment to its crisis between rounds of shuffleboard, or in the lounge over tea, or in reclining chairs on the north side of the boat-deck out of the fierce heat of the sun.

It was the attractiveness of these people that seems to have surprised and impressed him. He had of course lived a sheltered life, and he was not to know that men who drink and play cards on the Sabbath are not therefore all blackguards. He had probably known very few ungodly people: his parents and their friends, his sisters, the college authorities, and his fellow students had all been believers and active Christians. He knew too little of the other world not to be surprised to find that it contained people with their own pleasant virtues. He found some of them to be extraordinarily congenial. And perhaps it is here that his affection for the ungodly began to develop. Much later when he, W. E. Sangster, and Donald Soper were all ministers in London, drawing the largest congregations in England, someone made a summary of their dispositions as follows: Doctor Sangster loves *Je*sus (he used to emphasise the word so); Leslie Weatherhead loves his people; and Donald Soper loves an argument!

Well, you could say that. Part of Les's genius unquestionably resided in his affection for people—all manner of people. Like other members of the cloth, no doubt, he was always slightly on the defensive when it was argued that pagans were just as pleasant as pious Christians. But he himself seemed to love the pagans as much as the pious, perhaps more. And I think his ministry tended through the years to be directed less towards the official Christian, the signed-up, regular member, so to speak, and more to people he called "lovable pagans"; so that his last large book was written for the man who stood detached from traditional dogmas and pieties and could be designated by the words of the title, the term Les used also for himself, *The Christian Agnostic*. "I have always been attracted," he writes in the preface, "by those lovable men and women who rarely have anything to do with organised churches but who would never act dishonourably or meanly, who are full of generosity and helpfulness if ever one is in trouble or need, who bear their own troubles with magnificent courage, who never complain or grumble or gossip or run other people down. I want to write for them." People like this are, of course, pretty rare, though there may be hundreds of thousands or so in the world who approach the model. It seems to have been some of this kind whom Les met on the steamer and got to know for the first time in his life. In later days he would have been more urbane than to have put to them the disturbing questions that must have arrested conversation over the tea-things. But then in those early days he was a knight of faith, a young crusader, full of zeal, willing if necessary to be dismounted in the cause.

They reached Colombo at dawn on November 13th. With some fellow missionaries, Leslie spent a day or two there, and then by train, boat, and train journeyed to Madras. He had expected to be stationed here, but for a few months he was appointed as assistant at a church in St. Thomas Mount, a small village some miles outside the city. To this place, according to tradition, the apostle Thomas had come at the end of his days: he had founded the Christian churches on the south-west coast of India, had crossed over to the east, and then, near the mountain that now bears his name, he had been martyred. For Les, this was the beginning of his missionary work. He lived in a fine big bungalow with the minister he was to assist; he notes in his diary that the minister and his wife were "simply topping" and adds, "I ought to be very happy."

He was not, however. He was homesick for his friends and for England. There were not many English people around; and at this stage, having only a few words of Tamil, he was cut off from any commerce with the natives. He found his work uncongenial. He tried to make his new foreign den a home from home and began to embellish the walls with the coloured photographs from English magazines. But in the evenings, as he sat alone reading by oil lamp, strange exotic bugs flew in through the bars of the unglazed windows, lizards crawled up the walls and stopped and stared at him, enormous spiders scuttled along the wainscot, foreign noises floated in from the jungle, and from ponds nearby the frogs made the night loud with their croaking. His host and hostess regularly retired early to bed and left him to his own devices. He confides his miseries to the pages of his diary, and there he faces with some horror the possibility that his coming to India had all been an awful mistake.

The homesickness and the sense of alienation were to remain with him for some time. But things looked up when, early in the new year, he was moved to Madras to become the minister of Georgetown Church. Madras was then a city of about half a million souls, most of them Hindu. It was in no way a city of any striking beauty or nobility: one densely populated sector, the business district, known for many years as Blacktown, had recently been renamed Georgetown, after a visit by George, Prince of Wales. And here on Broadway, which was in fact a rather narrow street, stood the Georgetown Church, the oldest Wesleyan church in India. It was a heavy stone building with four Ionic columns and a pediment in front; surmounting the scrollwork of the capitals were the words, "Wesleyan Chapel 1822 & 1844", in large plain block letters. The whole edifice was imposing—for our taste, perhaps, too imposing, built, it seemed, to last forever, by men of a single faith in their own culture and complete confidence in their mission in India. Although the building published the date of 1822, worship had actually begun to be conducted on that site five years earlier, in 1817, and thus shortly after Les succeeded to the pulpit, the church celebrated its centenary.

He played no prominent role in the centenary services, which were conducted in March 1917. As the date drew near, however, we find him preoccupied with the impressive number of conversions that were being registered. In Wesleyan churches it would often be the practice at the end of the service for the minister to call upon those who had been converted and ask them to stand up to signify their decision. Les had engaged in this practice back in Farnham when he began his ministry, putting to the younger members of the crowd the challenge of declaring their faith. He set great store by the numbers of conversions obtained after the service, praising God when these were substantial and blaming himself when they were not. At one point in the diary four pages are devoted to a list of recorded conversions dating from the beginning of his ministry until the time he left Georgetown for the army. There are the names of the people, the name of the church where the transaction occurred, and the title of the sermon or the text that had brought these men and women to this point and had brought them one by one to push aside the prayer hassocks and come shuffling awkwardly to their feet, among the flowered hats and the frock-coats of a church Sunday, with some soft largo breathing from the organ. Thus, for example, the diary notes five ladies converted on February 8th, 1914, back in England at Plumstead Common, after a sermon on "Lo I am with you"; and on April 30th, 1916, an army private registered his commitment, following a sermon on Revelations 13.8, which is a rather harsh either/or prognostication—"And all that dwell upon the earth shall worship Him, whose names are not written in the book of life of the Lamb slain from the foundation of the world"—a sufficiently cryptic utterance, which must have called for some skill in exegesis to bring its message alive and warm to a congregation in the Home Counties.

On February 25th, 1917, in the Georgetown church, Leslie preached on the text, "What shall it profit a man if he gain the whole world and lose his own soul", and afterwards, responding to his call, twenty-two people stood up to be counted. The diary records some exultation. "Glory to God," writes Les, "for His wondrous outpouring . . . How gloriously God has worked tonight. What an answer to prayer." Two weeks later he preached on "What think ye of Christ?" and thirteen more souls were saved. The diary tots up the figures—"That is thirty-five since I came"—and declares that they are going to work and pray for a hundred to mark the centenary.

Although he always attributed the credit for these conversions to the Lord and His grace, Les must himself, as the chosen vessel, have received no small measure of joy. His life was a bit happier after he moved to the Georgetown church. He is still homesick from time to time, however, and he misses his friends. A sense of the appalling foreignness of India, coupled with some repugnance for it, still overwhelms him occasionally. But years later he remembered these days as happy. He lived at Egmore, a suburb of Madras, in a compound which contained the bungalow of his host, the Egmore Wesleyan Church, a tennis court, beds of red canna lilies, and some coconut palms. Les would be woken at dawn by a servant bringing "chota", a meal of tea, eggs and fruit. He would study until noon, when "tiffin" would be served, a large meal, often a curry with all its proper embellishments. An Indian curry is a hot dark ochreous food that sears the mouth and is not to be lightly or frivolously entered into. The full-dress version may contain onion, garlic, butter, green pepper, broth, milk, rice flour, coconut milk, blanched almonds, sultanas, green ginger, currant jelly, tomatoes, thyme, parsley, and curry powder, along with whatever meat, fish, shellfish, poultry, and vegetables are to hand, served with saffron and chopped up bacon, in a ring of seasoned rice, with mango chutney and bananas to cool the scalded palate. In Egmore, this would constitute tiffin; and after it was over, everybody went to rest, and the servants closed up the bungalow and went home. They appeared again at 3.30, to keep the wolf from the door with a light meal of tea, sandwiches, and cakes.

After tea, Les would go off on his bicycle to visit church people who were sick or in some other distress. Such visits would take him often enough into the slums of Madras, where he was stunned by the vile conditions under which parts of Indian humanity existed, and his Christian mettle was sometimes sharply tested in overcoming the natural repulsion of the senses. It must have been good to return after such duties to the compound, where there was sometimes a chance of a game of tennis before dinner. And then afterwards, books, gramophone records, walks in the moonlight, or bridge.

But to return to the compound after visiting the sick and the poor was not entirely to leave all the other behind. On some occasions the horrors he had witnessed stayed with him. Once in India in the hills he visited a leper colony and talked with the victims of this dreadful disease, which was then virtually incurable. Les was an extraordinarily sensitive man, and one can imagine him recoiling from the shocking sight—the white ulcerated skins, the hairless heads, the swollen lion-like faces. He felt something too of the loneliness and the mental anguish of these outcasts. It was good to come out of that settlement. But the images remained, along with others of men and women in pain, and so did a sense of irresponsibility—the feeling that one was able to turn around and leave seemed somehow culpable. Years later, in his book *Discipleship*, it was to provide an image for a concept of Christ:

> Sometimes [Les wrote] on some errand of love I go to the slums or some foul den of thieves or some haunt of evil. I pity. I sympathize. I

try to help . . . Then I come home. I bathe. I eat. I sleep. *But He stays there* . . . How well I remember visiting a leper settlement in India, speaking to the lepers, spending the day with them, and then—having sympathized and done any small thing or said any small word to help—I went to my bungalow, changed my clothes and sat down to dinner. But when Jesus saw a leper He became a leper . . . He identified Himself with all the horror of it until He could get His shoulder under the burden and lift and dispel it for ever.

There were good days for Les in Madras, for all the shared burdens of others, until shadows darker than these began to fall across them. About the time that he had arrived in India, the Battle of the Somme had been concluded at the cost of 400,000 British casualties; by December 1916, Rumania had collapsed; in the new year, Russia was on the brink of collapse; in February, Germany had initiated its unrestricted submarine warfare, which by early spring had become so effective that the end of British endurance could be foreseen. In addition to the untold expense of human lives, the war was costing over five million pounds a day. Such were the items of news that were reprinted in the *Madras Times* and brought each day to the bungalow through the exotic gardens of the Egmore compound.

On the first day of 1917 Les had solemnly recorded in his diary, "I am not doing my maximum for my country in her hour of need", and noted that he had applied for a chaplaincy in Mesopotamia. There was no vacancy, but having made the effort he seems to have appeased his conscience for a bit. By spring, however, he was again restless, and in early May, having passed a medical examination less searching than his earlier one in England, he applied for a commission in the Indian Army Reserve of Officers. The Missionary Society refused him permission to join the army, and, coming closer to him, his father sent him a cable telling him to be guided by the Society. He learned that he might be expelled from the ministry for disobedience, and he remarks in his diary—somewhat histrionically—"This would be awful, by far the biggest tragedy of my life." But not less now than at all other important moments of decision, he gave ear not to the voice of external authority but to the inner voice of conscience. He was, after all, a Protestant.

He was gazetted in July and in mid-August left Madras for Sabathu in northern India. By this time he had gained quite a following among the church-goers of the city. One Sunday evening in August when he was preaching, the Georgetown church was filled to capacity and a number of people had to be turned away from the doors. On his last Sunday, there were demonstrations of affection which moved him deeply. And then the following day a crowd gathered at the railway station and sang, "For he's a jolly good fellow", to the young Lieutenant Weatherhead in the brand-new khaki uniform with its shiny badges of rank, as he stood at the window of the train trying to contain strong feelings in a fragile grin.

Slowly the train pulled out on its long haul to Bombay, and Les began to make the acquaintance of his two fellow passengers, both subalterns. He records that one was "a very jolly goodhearted fellow though without Christ". The other, however, he says he "can't understand. He swears and drinks and boasts and openly says he only joined up on compulsion." This man seems to have belonged to that broad stratum of social life with which Les was not yet familiar, though he was very shortly to become so. Nor at this point, apparently, had he any idea of what important factors swearing, drinking, and conscription were in the defence of the country in her hour of need. He himself very rarely swore. He used euphemisms like "dash it", or "bust it", or "flopping" rather than the oaths or obscenities themselves. He very occasionally used a bad word in the service of humour or quoting a joke—but then, as will appear often enough in these pages, a joke to Les was worth buying at practically any price. In the train to Bombay he was gratified to observe that when his companions discovered he was a minister they let up a bit on the foul language. It need hardly be said that Les got on well with both of them: and this was good, for they were sharing a five-day journey.

Their way led across India to the west. They had left Madras on Monday and they arrived in Bombay on the Wednesday morning. There they got the Punjab mail train and went north to Agra, Delhi, and Ambala. From Ambala they went up into the foothills of the Himalayas by hill trains. And then for the last ten miles of the journey they travelled along mountain roads and ledges in a *tonga*, a light two-wheeled horse-drawn cart, which deposited them finally at Sabathu.

Sabathu was a hill station where infantry officers were trained. From the window of his room Les could see snow-clad peaks in the north; and throughout his time in the camp, whenever routine permitted, he would stare at these peaks and study their changes as the light changed and the heat, and as clouds moved across them or descended and concealed them. One peak (which he doesn't name in the diary) he singled out for his special attentions. And years later when he wrote about the Transfiguration, the mountain he had in mind and described was not Mount Tabor in the plain of the Jordan valley, but one of the Milayas range in Kashmiri as it had revealed itself to a meditative young officer in an infantry-training barrack, dreaming of some reality that might transcend the rigorous and squalid daily routines.

For it was a hard life. There were route marches in full kit, drills, rifle and machine-gun training. The officer-recruits were in the charge of non-commissioned officers, and these had to salute their trainees and call them sir. But they themselves were the real soldiers, and on parade-ground and march they could make the young gentlemen sweat. Route marches in mountainous country would probably be hard going for anybody: the air is thin at those heights and it takes some time to get used to the lack of oxygen. Les was not robust, and when he returned to camp from a march or from drill he would collapse on the floor of his room and lie there for fifteen minutes or more, too exhausted to unstrap his impedimenta. The instruction was accompanied by biting sarcasm prodigally laced with obscenities. Les learned not only the present arms and the butt-salute but, along with them as a bonus, a lot more of the new vocabulary he had been introduced to on the train.

Not only the n.c.o.'s but most of Leslie's fellow officer-students were the kind of people he was not used to meeting. He was still disgusted and even shocked at the swearing and drinking and the kind of orgy that passed in such circles for jolly good fun. One drunken evening in the mess, men walked over the dinner tables singing dirty songs and one of them climbed into the rafters to pour whisky on those below. Another kissed the native manservant and waltzed him around the room. During the course of the evening, fifty glasses were broken. On another occasion the Church of England chaplain, under the influence of liquor, sang a bawdy song, waving his glass to the music.

Les got pretty tense about this sort of thing, and we can imagine him sizing up these young imperial sahibs and ruminating about the irony of the master race bringing civilisation to the "lesser breeds without the law". What was most immediately distressing, though, was that this kind of performance separated him from his fellow men; and he liked his fellow men. He didn't think these people were vicious. He never, I think, possessed a strong sense of the evil in men. He was not backward in seeing stupidity and absurdity, nor, certainly, on the other hand, the capacity for courage and nobility that men manifest in their suffering. But he was slow and even unwilling to recognise wickedness. It belonged, no doubt, to his calling to deny that there were men totally given over to it; but along with his friendly weakness for liking practically everybody, he seems to have fostered, even in those early days, a belief in moral relativity which made him unwilling to locate evil absolutely. It was the product, partly no doubt of rationalism and modesty—who was he, he felt, to dictate to others what was right and wrong? But it ran deeper: it belonged to his liberalism and an unwavering faith in human nature. It was characteristic that in his book *Personalities of the Passion* (1942) he was to argue that Peter had not betrayed Christ and that Judas was not a traitor.

Some months after the Sabathu days, wandering in the deserts of Mesopotamia on his own, thinking on matters of right and wrong, he determined to re-assess his own attitudes to some of the conventional values of the Protestant churches. (In the same diary entry he notes quite incidentally that somewhere near in this desert is the site of Sodom and Gomorrah; and our veneration need not preclude us from a little amusement at the thought of the great evil of these cities and the horrifying retribution that paid for it as a background for Leslie's reappraisal of the middle-class virtues and prohibitions—whether drinking, or smoking, or going to the flicks were to be countenanced by a Christian!) A little later, he rehearses in his diary some of the principles of pastoral care, and he is liberal: "Let not the minister say, 'This is wrong'," he writes. "It is a question of individual conscience . . . Right and wrong are matters of degree." He always held to St. Augustine's rule: "Love God and do what you like", believing that the first exhortation controls the practice of the second. The liberalism lasted for many years. It became a habit with him that he could hardly find it in his heart to denounce a sinner, but would rather search out the grounds for sympathy. "Poor chap," was a more frequent response on his lips than "Scoundrel".

All the same, from such entertainments as diverted the young officers and gentlemen in the Sabathu barracks, he had quite frequently to stand aloof. There were many situations, in fact, that alienated him. One day when his platoon was waiting to enter the firing range for the first time, the men organised a sweepstake on the shooting. Les would not bet. He was charged with being superior and snobbish; and one officer, who knew he was a minister, scored an inexpensive laugh by observing, "Parsons can't shoot, in any case." Les was saddened by the gratuitous hostility. But on this occasion, this parson proved both militant and triumphant: at a hundred yards, he got five bulls'-eyes out of five; he got a full score in snapshooting; and again five bulls out of five with the machine gun. The gent with the misgivings about clerical fire power was surprised and a bit abashed. "You could've made twenty rupees," he muttered. Les was surprised himself.

On Sundays there was no drill, and the officers were at liberty to do what they wanted. There was not a lot to do in Sabathu and some of the men would hire a *tonga* and go off to Simla. Others would organise tennis or cricket, or they would study the military handbooks that were required texts for the course. Leslie's practice was to walk off into the hills with his Bible and hymn book and read passages of these, out in the lonely places away from the barracks and in scenery that meant much to him, not only for its visual splendour but also for the fact that on those high snow pastures no human foot had trod, and in the shadowy valleys among them no human eye, even, had ever peered. He felt that these landscapes were very particularly the Creator's own, and these external glories nourished his inward life.

How characteristic of him is this instinct to walk away from the noise of human affairs into the wilds, where the spiritual dimensions of life seemed to him to be more clearly revealed! In the Moffat hills, in the woods around Farnham, in the foothills of the Himalayas, later in the empty deserts of Mesopotamia, and even at last in the municipal park in the Borough of Finchley, London N.3, we find him continually withdrawing from a world too much with him, never less lonely than when alone, always looking for silence in which to restore depleted powers.

In October 1917, at the end of training, he was posted to the 88th Carnatic Infantry. He had applied for this regiment because it had a detachment in Palestine. But it also had detachments in Madras and Mesopotamia, and it was to St. Thomas's Mount, Madras, that Lieutenant Weatherhead was ordered to report. The divinity that shapes our ends must have seemed to him unusually irrational when, the following month, he found himself preaching again to a crowded congregation at the Georgetown church, drawing his old friendly crowd and duly recording conversions.

He spent Christmas and New Year in Madras. But then in early February of 1918 he was shipped out of Bombay for Basra, in Mesopotamia, the port at the head of the Persian Gulf. After about a week at sea they lay at anchor in the river which, according to tradition, had flowed through the Garden of Eden. When he had been in Basra a day or two, Les, never backward in arriving at his own heretical conclusions, whimsically reformulated the doctrine of the *felix culpa*: the Fall had been fortunate, he decided, if only because it had delivered our first parents from this flat desolate area where nights were bitter with frosts and rain and cold wind, and days insupportable with the dust and the glare of the sun.

He was attached to the 83rd Wallajabad Light Infantry which was encamped in the desert on the east bank of the river, guarding Basra. It took him a little time to get to like the life: the camp was ankle-deep in mud and the wind coming from the north brought the cold into his leaking tent. The jackals howled in the night. Sometimes Arabs in the uniforms of British soldiers would infiltrate the camp and steal weapons. But before long, Les began to get used to the hardships and to find things to enjoy. He was issued a horse, a thoroughbred Arab, which he named Ginger after his younger sister, for it was, he writes, "splendidly affectionate". He used to gallop out into the desert to watch the most spectacular sunsets he had ever seen. He got to love the starlight of the desert and the silence which, after the sounds of day had died away, became so intense that it rang in his ears like a bell. Most of all, however, he got to like his men. They were Indian soldiers and he taught them to play hockey. They liked him in return, and one of the officers overheard them say that if they had to go into action they would like to be led by "Weatherhead Sahib", for then they would all get the V.C.

It was not to be like this, however. In March 1918 Weatherhead Sahib was invited to become an army chaplain. It seemed to him that in this role he could meet the demands of the two major duties he had laid upon himself—to serve the country and to serve God. He had not been eager to be a chaplain; he had felt that they did not fully share the life of the men and that they ought at least to have had a baptism of fire. But now the major fighting in Mesopotamia was over, and it looked to Les as if he would be in that theatre for the duration. Thus he told the Senior Chaplain of the area that he would transfer to the Chaplaincy Department if needed, and the wheels of the massive army machinery were accordingly set in motion. Some six or seven weeks later Les received a phone call from the Senior Chaplain to the effect that his transfer had gone through G.H.Q.

However, the inscrutable ways of the divinity in shaping ends were once again seconded by the even greater inscrutability of the army: within an hour of learning that he was to be a chaplain, Les received new and contrary orders from his C.O., on account of which he found himself a few days later in command of Paddle Steamer No. 22, which bore him, along with a hundred and fifty Arabs, a handful of British and Indian soldiers, and a lot of cargo, towards Amarah, about a hundred and twenty miles up the Tigris from Basra. All he knew of Amarah was that it was located in a green part of the relief map. He didn't know exactly what he was supposed to be going to do there when he arrived, and he writes in his diary that it was a "special stunt" associated with the Department of Local Resources—"probably . . . something about which I know nothing", he remarks. But the whole idea of the adventure appealed to him greatly: the boy who had been stirred by Henty, Ballantyne, and Rider Haggard in the back bedroom in Leicester was now surely realising his dreams. The thought of the unknown, he writes, made his blood tingle. Meanwhile, as the paddle-wheels churned up the yellow river mud and the Arabs chanted their tuneless music in the bows of the boat, this most unconventional adventurer betook himself to a corner of the bridge out of the sun, pulled out his pocket hymn-book and read aloud to himself, "In Heavenly love abiding/No change my heart shall fear."

After Amarah he went on some eighty or ninety miles upstream to Ali Gharbi, where he was ordered to take soldiers and explore a tributary of the Tigris. When he had made his preparations, however, this expedition was cancelled, army fashion, and he was ordered to take a group of British soldiers down the river. These men had been selected for their knowledge of farming, and Les deposited a few of them with each of certain Arab sheikhs on the left bank of the river. They were to encourage the Arabs to grow grass and corn, to reap the grass on the river banks and make hay, and to irrigate the land. Hay had not previously been made in this area, and the British needed it for the horses, for which fodder was being shipped from Bombay. When he had delivered his men, Les rode on horseback from one encampment to another with his interpreter, a Baghdad Jew who spoke English, French, Arabic, and Hebrew. They had to pay for the hay, and to make notes on the terrain, some of which had not previously been explored. At the same time he was engaged by the Political Department of the army to make friends with the sheikhs and improve, if he could, their attitude to the British.

At some places friction had developed between the British soldier-farmers and their hosts, which was hardly unexpected, since the Arabs had no English and the British no Arabic, and all transactions had to be made by sign language. At one place the stack of hay the Tommies had made had been pulled down by the Arabs, and Les had to investigate the quarrel, arbitrate, and get the hay set up again. There were sharper hostilities than this, however. The Turks had withdrawn from Mesopotamia about a year earlier, but there were still a number of sheikhs who remained loyal to their recent colonists and were unfriendly to the British. One night, at the banquet, as he squatted on a cushion on the floor among his hosts, eating after the Arab fashion—the left hand behind the back, the right hand extended into the common dish of chicken, rice, and eggs—Les heard his interpreter whisper, "These men mean to kill you." After the meal he contacted the soldiers, and they saddled their horses in the dark and led them silently out of the camp. There was no moon. Outside the camp, they mounted and galloped off into the desert. They got away safely, but they were in unmapped territory, and before they had gone a mile they had ridden into a river, a tributary of the Tigris which had not previously been located. More than somewhat bedraggled they made their way to the next encampment, but one of the soldiers developed a fever, and they had to flag down a boat on the Tigris and have him shipped back to the hospital in Amarah.

Les's encounters with the tribesmen were not usually so sensational. Sometimes their customs were irksome, though: he became tired, for example, of the unrelieved diet of chicken, goat, and rice, and he and his men longed for the habitually despised cans of "bully beef", as supplied by the army, a meat reputedly derived from the cheaper quarters of old and tired animals. Also they lived in almost continual thirst: having endured the heat of the day, having sometimes been many hours in the saddle, they were tantalised by tiny cups of goats' milk—the only available alternative to the flat tepid water of the Tigris. The Tommies made light of these privations, however, and their resourcefulness amused Les endlessly. He saw two of them squatting on the ground over a flypaper that had been divided into equal parts, betting on whose half would catch the most flies. At night he heard them chattering in their tents, and by now he could be amused by their exorbitant use of obscenities. And then during occasions when their hosts were present, the Tommies would discuss them quite openly, assuming correctly that their language would be incomprehensible to the Bedouins—as indeed such English dialects are often incomprehensible outside their own counties. "Cor, look at vat bloke," one of them once remarked, as an Arab sheikh lowered himself down on to the cushion for the banquet. "Ain't 'arf a fatty, in't 'e? Look at the pot on 'im! Bet 'e ain't done a dy's work in 'is lahf."

Sometimes the courtesies and the hospitality of the Arabs were overwhelming. We have noted that one of Les's most repeatedly exercised instincts was to move away from the madding crowd in order to be alone to think and meditate. One day, when he was staying with a Sheikh Banayeh, he wandered away from the encampment and out into the desert alone, only to find to his annoyance that he was being followed. When at length he stopped and sat down, Arabs came up with cushions, carpets, and silks. And as he wrote his meditations in his book, the sheikh himself came out of the encampment and ensconced himself at Les's elbow, determined that this young western officer should not lack any conceivable courtesy that he might proffer.

Towards the middle of May of that year Les stayed for some two weeks with Sheikh Belasim in a camp which lay farther away from his base than he had hitherto ranged. When he had first arrived, dismounting half a mile from the camp and leading his horse in, as demanded by the traditional custom (breach of which might mean being shot on sight), the sheikh had walked out to meet him. During the ceremonious encounter, the sheikh's fourteen-year-old son climbed on Les's horse and in a moment was thrown, the pommel of the saddle tearing a gash in his stomach. Les cleaned the wound and bandaged it, and the boy subsequently healed. Belasim was tremendously impressed, not only at the medical skill but also at the fact that the English officer should *himself* kneel on the ground and dress the wound.

But Les was not apparently the first westerner to have visited these parts. The Arabs showed him a golf ball they had found, and they brought it to him in their finger tips, suspecting it might be a bomb. He explained what it was, feeling that the western way of life must seem a bit curious, as he told them that this was a thing that men in England tried to hit across a field and into a hole! Other aspects of the western way must also have seemed curious to the Arabs. Belasim treated his guest with great courtesy. He gave him a tame gazelle. He organised horse races, and when Les won he made him a present of the Arab pony he had ridden. One night as Les sat before the door of his tent, he saw a glimmer of white approaching, which proved to be the gauzy white robe of a young girl. An Arab held her lightly by the elbow, his other hand extended, palm up, offering her. Her eyes were lowered and her hands hung by her sides. Les shook his head.

The Arab shrugged lightly and returned into the darkness. Shortly the glimmer returned: now Sheikh Belasim himself propelled the girl forward. Les saw difficulties arising as the oriental courtesies came up against the gritty resistance of British morality; after all, he was supposed to be fostering good relations up and down the river, so he sent for his interpreter. "Thank the Arabian Lord profusely," he instructed him, "and explain that the British officer prefers to sleep alone." Belasim uttered a voluble protest. "He says, Sir, that she is a virgin." Another protest followed: "She would be honoured to sleep with you, he says." "Tell him," said Les, recalling the images from an Arab love-song he had once heard, "tell him that she has the soft eyes of a dove, her hair is of silken threads, and her skin is as soft as the skin of a pomegranate, and it is not on account of any failing in her beauty that the Arab lord's lavish generosity is declined. Tell him all that," he said, "and then explain that it's just part of our curious British way of life." To his relief he saw that the sheikh had got the message: as the interpreter concluded the piece, Belasim threw back his great head and rocked with laughter.

There was clearly no resentment. He withdrew again into the dusk, the girl by his side; and then within a few minutes a servant appeared carrying on his shoulder what turned out to be a huge Arab carpet, woven in turkey reds and golds and greens and brown, a gift not less exotic but certainly more durable as a souvenir. It lay on my father's study floor for fifty years, "Just to think how I came by that carpet," he used to say. Later, much later, when he repeated the anecdote, we would ask, "Les, which would you really rather? Come on, now!" "The carpet," he would snap. "You're wicked boys. Besides," he added, "some of the Devonshires who were with me there turned up again later back in the regiment when I was the chaplain."

Leslie stayed long enough with Belasim to be able to engage in some of the local pastimes, such as jackal hunting and horse racing. He also kept a few pets, beside the gazelle—a pariah puppy, a lamb, and a young jackal. He swam in the cool water of a nearby *wadi*. Once again, as he does not fail to mention in his diary, there were snow-capped mountains in sight, when he looked eastward over the desert towards Persia; and they were "a never failing joy".

He seems to have enjoyed life pretty well among the Bedouins. The worst of it all was probably the insect life: in the night, moths, fleas, flies and beetles shared his quarters, some coming surreptitiously, others driving down in herds whenever he lit a lantern. In the end it was the insects that brought an end to his activities in the desert. He spent much time in the saddle wearing only shorts and a shirt, and gradually the leather pistol holster on the Arab saddle chafed his knee. Before he could get the sore to heal, it became infected, and he developed sandfly fever. He made his way to the Tigris, stopped a steamer, and consigned himself to the British hospital in Amarah.

He was there two weeks, and it was fine to be in the western setting once again, with a white-sheeted bed, the clean white plumbing, the variety of food, the long drinks of cold water, and the starched young nurses. Unfortunately his ailment came to an abrupt and untimely end and he was summarily cured and discharged. Among the men in the ward, one of the leading diversions was cheeking the nurses, their favourite victim being a young Scotch redhead, who responded in kind and gave as good as she got. Once, as she came into the ward Les flung a pillow at her head, in order to help keep up the morale of the men. The girl must have ducked, though, because what he hit was the purple face of the senior medical officer, who thereupon readjusted his spectacles, gave the temperature chart a perfunctory glance, and told Les he was well—a reasonable declaration.

In the British army if you can't hit a red-head with a pillow at fifteen feet, they let you become a chaplain. At least, it seems to have been at this point that Les's transfer came through and his service as chaplain began. It was to continue from July 1918 to April 1919. During these months he moved about Mesopotamia from one army outfit to another, and on the whole his experiences seem to have been unhappy. For one thing, his health, never robust, repeatedly failed him: he had Spanish flu, English flu, and a miserable series of boils. The climate was tough on him: that summer the temperature rose sometimes to 120 degrees at midday—a heat hardly to be borne even in tall rooms with waving fans, and practically insupportable in a small tent. Worse than the physical debility and the heat, however, was the hostility or at best the apathy of his fellow officers that Les frequently encountered in the camps where he was stationed. In one place, he writes in his diary of his words being maliciously twisted and of the distrust and suspicion in which some of his fellows held him. In another camp, the commanding officer announced in the mess that chaplains were worthless and a mere nuisance in the regiment. "I can read the service myself," he argued. Les notes in his diary that the colonel was something less than a gentleman: but at the same time, it is clear from the meditations that accompany his sporadic observations that he held himself much to blame for the attitude of others: if he himself had been an exemplary Christian, he believed, men would not have despised him for his religion. A few years later, in an article which became the first chapter of his book, *The Transforming Friendship*, he writes, "It has been truly said that Jesus could go into any Army mess, into any factory dining-hall, into any business or professional common-room, into any hotel or boarding-house, into any students' hostel or college, and His presence would not make men uncomfortable." True perhaps; but an example of which even the very best would fall short.

I think Les suffered rather extensively from feelings of guilt and unworthiness. There was nothing morbid about them: he was never the gloom-bitten puritan wasted by sin and conscience. Quite the contrary: he was to all appearances as jolly as a publican; and cheerfulness, he thought, was a duty. Beneath it, though, he sometimes thought himself scarcely fit to speak of Christ and to recommend the Christian way of life. Often enough in the diary, back in the early days of the century, he writes words like these: "I am unworthy to be called by Thy name, let alone work for Thee, but I plead Thy love and Thy cross, and I know Thou wilt not turn me away." He felt like this regularly, I'm sure. I remember when, nearly half a century later, Macmillan's government was shaken by a sex scandal, and Les, then at the height of his fame,

Wedding picture, 1920.

Les in 1940.

wrote to the protagonist to say—probably quite untruthfully—that in the sight of God he himself had done worse things than those which were now a sensation in the press, but that he had never been found out. Well, I think he always had this feeling that he wasn't as good as he looked. And it was like him to discredit his own image to give encouragement to another.

It is not possible to imagine that, when they got to know him well, his fellow officers and men could easily sustain their animosity. Nor may one suppose that his preaching abilities were without effect on those who witnessed them. At one camp, at least, he notes that, after he had gone to some considerable effort to organise a church service, only four men showed up out of a battalion of eight hundred. Six weeks later he notes that services are "splendidly attended" and the church tent is packed on Sunday nights. But as he made friends and, indeed, converts in one camp he would be moved on to another and have to start again.

He began his chaplaincy at Nasiria, on the Euphrates, about a hundred and fifty miles from Basra. Early in October he was appointed to Basra Mobile Column, ministering to a Devonshire regiment, among other units. Later in the same month he was moved to Ashar, near Basra. Dusk came early in the winter there, and fog moved off the river and crept up the narrow streets. But there was one memorable day when all the boats on the Tigris blew their horns and sirens and rang their bells when, after a month or more of rumours, peace was at last declared. In November Les went up the river to Bakubah, near Baghdad, to join the first fourth Devon regiment; and he was stationed here from November until April 1919 when he was released from the service. During this time, he began to enjoy life a little more. There is a snapshot of him in uniform, wearing puttees and a curious knee-length sheepskin coat, cut off at the elbows. He is standing in front of his tent, grinning, a cigarette stuck in his mouth at a jaunty angle. Beside him are the adjutant and the quartermaster. Behind the tent, the lone and level sands stretch far away. He looks pretty happy. In Bakubah, once again he began to draw large congregations to the church tent. He organised a choir and a football team, debates, and lectures. He rode much and hunted jackal by moonlight. Here some of the men were detailed off to administer living facilities and hospitals for 35,000 Armenian refugees, whose beautiful women bathed naked in the rivers on ceremonial days.

It was while he was at Bakubah that Les had the great excitement of a trip to the north-east, over the border into Persia and almost up to the Caspian Sea. He, another chaplain, and an orderly took a Ford pickup up over the high arid mountains and down through wadis in Kermanshah,

Hamadan, and Kazvin. The territory had been successively laid waste by Turkish and Russian armies, and many of the villages had been deserted by all but a few hungry people, to whom they gave what provisions they could spare. The land was infested by brigands. There is a road now marked red on the map; but when these three went, they were told (unlikely though it seems) that they were the first to take an automobile that way. Their ostensible purpose was to estimate the need for chaplains with the British units placed throughout that area. But it was probably a bit of a pretext: adding to the thrill of seeing new places was the legend that this was the route of the Three Wise Men, or part of it, as they came from the East to Bethlehem, with gifts.

Les was released from the army in April 1919 and made his way gradually back to Madras. He might at this point have claimed a passage home to England, following his own impulses as well as the recommendations of the doctor; but by offering to stay he was able to release an older man who had been out East for some six years and needed badly to return. Thus in the summer of 1919 Les became minister of the Egmore Church in Madras, in the compound of which he had lived during his earlier stay. In addition, he was superintendent of the whole circuit, and it was something to be a super at twenty-five. His ministry here continued until March 1922 when finally, for health reasons, he returned to England.

In the Egmore compound there were the coconut palms again, the canna lilies, the tennis, the siestas, and the good meals served by boys in white clothes: life went on as it had done before. Les did his visiting now, however, with a motor bike instead of the old push bike, arriving to exercise his pastoral care in a cloud of dust and a series of half-controlled explosions, a little oil-stained, but not so hot as before. The church congregations were larger than they had been earlier, and this was an encouraging tribute to his preaching, even though he wondered whether some of the word was not falling on stony ground as he observed that when, with the unexpected crowds the supply of hymn books ran out, the stewards took them off the natives and gave them to the whites. Not only were the congregations larger, however; there was also an increase in the number of men and women who stood up after the service to declare their conversion. In spite of occasional bouts of a strange fever that laid him low, life was pretty good to Les. And it was shortly to be materially improved.

Before he had left Madras, he had known a young woman called Evelyn Triggs—she appears in a list of friends he had drawn up in his diary, in some dark moment, no doubt, taking stock of his resources. Then, when he was in hospital in Mesopotamia, he sent her a card—

gratuitously, it seems, a mere friendly gesture. Now he met her again. She was the daughter of a Wesleyan minister who lived in the Isle of Wight, and, having come to Madras some years earlier as a missionary, she was now the vice-principal of a girls' boarding-school, Royapettah. The early photographs show the figure of a tall, slender girl, her hair gathered in a bun, her features simple and clear. She had a way of wrinkling her nose when she laughed, and Les called her *tungachi*, Tamil for little sister; she called him *Thumbi* (the *h* is silent) meaning little brother. But one evening there was a picnic on the beach, and after the others had packed up and left, these two lingered behind. It grew dark; a great moon rose and lit the little waves curling in from the Bay of Bengal; there was a light breeze in the casuarina trees; and it took them few words to agree that they wanted to move on from the brother-sister thing. They went back to Royapettah arm in arm; the prologues were over. That was in December 1919, and pretty soon Leslie was getting a sidecar attached to the motor bike.

Decisions are reputed to be hard for those born under the sign of Libra as Les was. And after they had become engaged he records in his diary how he is assailed by doubt: he quotes Tennyson, his old standby and comforter, "I falter where I firmly trod", deriving some comfort, as one does, from the expression of grief on the part of somebody else; and once again he seeks to distinguish his own will from God's: "If it is not Thy great divine purpose," he writes, "then break us both before it is too late."

But God did not, and they were married in the summer of 1920. The band of the 1/88 Carnatic Infantry, Les's old regiment, played "Berceuse de Jocelyn", and the bride wore a wreath of orange blossom and white heather. She looked "most charming", the *Madras Times* said, and added, "in spite of having spent five years in India".

They spent their honeymoon in Darjeeling in the north part of what is now Bhangladesh. The reader will not be surprised that Les should wish to go to the foothills of the Himalayas to within fifty miles of Kangchenjunga, a mountain of 28,000 feet. One morning they rose before it was light and took horses to a hill some miles away, waiting in the dark for the first ray of dawn to catch the peak of Everest. It was, one may well imagine, a breath-taking sight: many, many years later, Les remembered these minutes, which became increasingly magical, sacred, and symbolic for him as they became further separated in time: "Everest is so high," he wrote, "that it catches the light when all the world is still in darkness. Then gradually the light rolls down upon the lower peaks, then to the foothills, then to the jungle-covered valleys at the base, and finally to where we stood. We only saw Everest once, but it was unforgettable, like seeing a glimpse of the glory of God which shall one day be revealed to all mankind." Les was like that with mountains.

They settled down in Madras for a year and a half. During this time a trivial incident occurred which coloured, more or less literally, the rest of Leslie's life: he could never resist a bargain, and when he found a quart of green ink going at an absurdly cheap price it virtually determined the colour he was to use in pen and typewriter for over fifty years. He used a pen with a nib like a small spade which produced script in which u's, m's, n's, and, as he grew older, l's, t's and h's were indistinguishable.

A more significant and dearer acquisition of these months in Madras was their first child, Arthur Dixon, born in Vellore Hospital in the summer of 1921. Then, a few months after that, the doctors were telling Les he ought to move back to England for his health. The three of them left India in February 1922 and reached England some weeks later after a long and miserable voyage. It was snowing when they arrived. Les had been away for five and a half years.

3

Les was not due to assume his ministry in Manchester until September; so when my parents arrived in England in March of 1922 they went first to stay with my mother's parents, Arthur and Alice Triggs, in Ryde, Isle of Wight. Arthur Triggs was the minister of the Wesleyan Church. He was Cornish, a race of men with Celtic blood, more whimsical and visionary than their Anglo-Saxon neighbours. He was a shortish man, as I remember him, with a white beard and very little hair on his head. He had baggy trousers and walked with splay feet and was called Father Potatoes for uncertain reasons, but perhaps because he was keen on gardening. He had a greenhouse, and he grew tomatoes and geraniums, and there were brilliant dark red fuchsias under his study window. I think he was a saintly man, lovable, and very kind; and my mother adored him. He was a wit. Once when walking along a country road in the Island with my parents he stopped near a corner and bent down; and then, answering their curiosity, he pointed to the road sign: "It says 'Bend'," he explained. His wit was sometimes impish and caused annoyance to my grandmother; this, I fear, incited him to generate more of it. "I am A.T.," he would say, making his initials sound like eighty. She would frown at this. "And my wife is A.T. too (eighty-two)," he would then add. Mrs. Triggs was, in fact, older than him, and she was larger; she looked a little like a taller version of Queen Victoria in Her Majesty's later years. She bossed her husband around quite a bit, and needless to say she expected subservience from her daughter. Les soon recognised her as a fully-paid-up member of the Dominating Women's Club. When he spoke to Arthur, she would answer for him, until Les would finally say, "Let Arthur answer for himself." At this, Mrs. Triggs, quite undaunted, would continue to respond to his remarks, saying now, "Come along, Arthur, answer for yourself." We knew her as children, and she seemed a great dame; she had a great hat which rose tier upon tier. In the house in Ryde, I recall, there was one of those lofty toilets, thrones you could call them, such as only the British would construct and they only in their imperial days. It was a machine of some distinction and it had flowers in Wedgwood blue around the base and down at the water level inside. It was much higher than ours at home and I remember thinking that a more modest piece of equipment, however humble its function, would not be proper for my grandmother.

My mother and father stayed in Ryde for some time and then moved over to Cromer on the east coast, where Leslie's father had rented a house for all of them—Les, my mother and the baby, my mother's parents, and Les's two sisters and himself. All men occasionally entertain romantic ideals, and it must have occurred to him that having the two families together would be productive of jolly times. There were some tense ones though, Les told me: Alice, his sister, terribly sensitive to the idea of suffering in humans or animals (she later became matron of a home for cripples), protested when Mrs. Triggs cooked a live crab in boiling water. "They don't feel a thing," she assured Alice; "they die at once." The evidence, however, seemed otherwise when the beast pushed off the lid and hooked a claw over the edge of the pot. Mrs. Triggs clobbered it back in again and replaced the lid, but Alice left the room in tears. There were other tense moments too when the two sisters-in-law, maidens both, tried to instruct my mother in her management of the baby; Mrs Triggs also felt she should be heard from on this topic, and Les entered hotly into the engagement.

It must have been some relief for my parents to travel north to Manchester at the end of the summer and take possession of their own home. Manchester in 1922 was the centre of the cotton industry, a great city of some three-quarters of a million souls. It had some narrow streets and congested traffic. There was much rain, and the buildings were turned black by smoke. The church was in the university district, as was the manse, the house supplied for the minister during his tenure, supplementing his salary of two hundred and eighty pounds per year. It was in a shabby street of terrace houses, each attached to the two others on either side. One of the attached houses was a pub, The Lord Cecil; the house opposite was a brothel, where, on their first Christmas Eve in Manchester, they saw through the lighted windows a man knock a woman downstairs.

The manse itself was dark and shabby and the furnishings ill-appointed. House and furnishings being supplied, a Wesleyan minister sometimes had to endure some grotesque pieces for the duration of his tour of duty at a particular church. Such furnishings were often the products of bargain shopping, or pieces discarded by rich parishioners; at best they had been provided by someone whose taste was not that of the incumbent. It was sometimes possible to persuade the circuit steward, a layman charged with the care of the minister's housing, to change this or that. But often enough things stayed as they were—the chair with the broken spring, the sere curtains with the fading red peacocks, the wallpaper with the endlessly repeated flight of mallard ducks, the furniture with the marks of all the various occupants who had come and gone and left indelible scars, and the carpet, "a diary", as Les remarked in a letter to the *Methodist Times* describing these situations, "a diary of all the accidents that have happened in that room for twenty years".

61 Cecil Street, my parents' first English home, was unusually shabby: the linoleum was dark brown and cracked; there was dry rot in the passageway, a gas-pipe leaked in the cellar and fumes pervaded the whole house. The kitchen sink, a large stone piece, had settled at one end, and it was necessary to push the dirty dishwater uphill with a scrubbing brush to get it to go down the drain. But when my mother demonstrated this defect to the steward, he said, "But you seem to manage expertly," and the sink remained as it was. The letter Les published in the paper purported to have been written by a minister's mother-in-law to her husband and was signed Amelia Ray Tatwunce. It opened, "Any girl who marries a Wesleyan minister is either a heroine or a fool."

That winter, as the rain beat on the black houses and dripped down from the slate roofs in the dark street, Les must have looked back with some nostalgia to his previous home in India, with the tennis parties in the evenings or the walks in the moonlight with his new bride, or the long chairs on the verandah after dinner, and the white tussore-silk evening clothes, and the boys bringing cigars and cold lemon drinks. When his second son was born in the cramped upstairs bedroom in Cecil Street, Les rolled up the window blind and looking out into the rain said, "What a world to be born into!"

The church was a large unsightly building, seating a thousand people. For some weeks that autumn the Sunday congregations were sparse, thirty perhaps in the morning and forty at night; Les's carefully written sermons echoed among empty pews. One aged church member told him pointedly, "In my father's day, young man, the church was full, and the street was lined with carriages." In an amazingly short time, however, things were like that again. With some university students he knew, Les visited three thousand houses in Manchester to attract people to the church; each week he held discussion groups for students at his house, and at the end of the Sunday evening service he organised a social hour. Gradually word about his preaching spread through the circles of church-goers or potential ones. Once again the church galleries were opened on Sundays, and the street was lined with motor cars. He was shortly able to say to the aged church member, "This must remind you of your father's day," to which the old fellow replied, "Young man, don't be carried away by numbers. They are no sign of success."

But numbers there were, always—in Farnham, in Madras, now again in Manchester, and later in Leeds and in London, drawn by the extraordinary eloquence of his preaching—"a larger following probably," says Rupert Davies, in his book, *Methodism*, "than any other preacher of the age". And we may ask at this point what were the main factors in the manner and the matter of this preacher that brought men and women not only from every shire's end of England but also, in time, from the farthest corners of the English-speaking world, and also from Europe, Asia, and Africa, to hear him. There was, of course, the talent, the sheer gift, gratuitous and inscrutable, and then, matching it, a great deal of hard work, the background reading and the lengthy and arduous preparation of each sermon. At first he wrote out his sermons word by word, and in the early days, there were many that he wrote out three or four times. Later he prepared a card with headings and sub-headings in green and red ink, so that, while the structure was carefully determined, the actual phrases of the sermon were spontaneous, and he could look his congregation in the eye. He needed to have his notes with him. And once when Dick was driving him from Memphis, Tennessee, to preach in New Orleans, he realised after a hundred miles that he had forgotten them, and Dick, with a few patient and mild words of regret, turned the Buick round in the road and calmly drove him back again to get them. He wrote once, "Let the preacher face the people. He will know by their expression whether they are interested or bored or can't understand. Then he can react with fuller explanation. When I was a young minister a college tutor said, 'Prepare as if everything depended on you and then deliver fearlessly as if everything depended on God.'" He felt the importance of what he called the artistry of preaching, the relation of parts of the sermon to the whole, and the balance and the careful choice of illustrations for his theme.

Except for Monday, when he played golf, he worked each weekday morning in his study with his books around him, sometimes behind a huge oak rolltop desk but more often, as I remember him, sunk deep in a low armchair. He was not exactly incommunicado in the study, but if you burst in on him you might get a cool reception, and he would ostentatiously lay a finger under the line where he had stopped reading and give a mildly pained look over the tops of his glasses, a look which would let you know without any doubt that he had been interrupted and which would quote wordlessly, "When you take my time you take from me something I was intending to use"—wordlessly, because he would be too courteous actually to say such a thing and because the line comes from too modern a poet.

In addition to what was given and what was earned by hard work, there were two things, I think, that accounted in general for the wide and strong attraction of his preaching. First, he emphasised the humanity of Jesus rather than the divinity, which latter he regarded in any case as only an extension of the human spectrum; second, he emphasised the immediate day-to-day relevance of the gospel and not the minutiae of theology. In the sermons Jesus appeared repeatedly as a very good man. He was a man like other men, who loved the sunlight and the good air of the hills, the companionship of friends, food and drink, and rest after the day's work. He liked to be alone quite a bit. He liked children. He had a sense of humour. He was not superhuman, but superbly human: Les believed, for example, that Christ was not omniscient; He performed miracles, but they were not so much divine incursions upon natural law as the employment of powers which were available to the perfect human and, to a less degree, even to the imperfect: many of them could be accounted for and understood by reason and modern scientific knowledge, and they need not be classified as mysteries or magic.

He would point out that often there had been a passage of time between the acting of a miracle and the recording of it, and that the recorder had been influenced in the meantime by his own notions of divinity and associated magic. Thus he explains the feeding of the five thousand in naturalistic terms: a boy goes off to see the famed preacher, taking with him, for his lunch, sandwiches made from six flat loaves and three fish. However

> Boy-like, long before lunch, he eats one loaf and the whole of a fish. He presses as near to the Master as he can and overhears the conversation about the lack of food. Eagerly he presses on his Hero his remaining five loaves and two fishes. Jesus is moved by the gift. It mirrors His own spirit, and it suggests the solution of the situation. It is hard to believe that all these people came out into a desert place with no food amongst them at all. Jesus arranges for them to sit down in companies and in the spirit of the boy who has given his all, to pool all the food they have brought.

Someone remarked, reasonably enough, that if you're going to change the whole thrust of the story you might as well change a figure or two in the text and save the curious hypothetical arithmetic of the bit about the half-sandwich. But whatever its failings, Les's reconstruction succeeds in engaging a modern imagination.

He made the same kind of imaginative reconstruction of the story of the turning of water into wine at the wedding in Cana in Galilee. He did not deny that Christ *could* turn water into wine, but he believed that the vessels had been filled with pure water and that one of the guests, thrilled beyond measure by the incredible aura of the occasion and the personal charisma of Christ, had reported that even the water tasted like wine.

Often he gained a good deal of insight into the real lives behind the scriptures by appealing to the original Greek text and coming at the stories more closely than translations permitted. Many sermons came alive on account of vivid reconstructions of biblical incidents, and the key to these, often enough, was a knowledge of the Greek. He also knew the Middle East and was familiar with a number of the peculiar features of life recorded in the Bible: not only through the Bible and its exegetes, but also with his own eyes he had seen shepherds with flocks of sheep and goats, men working in vineyards and reaping fields, women coming to the well. He knew, of course, the classical scholarship on the Bible and he was a supporter of the Higher Criticism, which still in the 1920s engaged people in debate, but his sermons never sounded as though he had got things merely from books.

It was said of Socrates that he brought philosophy out of the clouds to dwell among men, and it might be said of Les that he brought theology out of the schools to be made relevant in people's lives. At the same time it would have to be added that one of his characteristic treatments of theology was to short circuit it. Quite often, to the horror of the orthodox, he would assert that doctrines such as that of the Virgin Birth and the Immaculate Conception simply did not matter. Jesus did not say, "You must believe the following articles"; He said, "Follow me." Ruminating about the preacher's duties some years earlier, Les had made the following note in a diary entry headed "Trinity Sunday, 1918. Aligharbi. Tigris Right Bank":

> Let a minister be practical above all things. Give a message on Sunday which will help people from Monday morning till Sat. night. Let him read Philosophy if he likes in his spare time but let him read books which will give him material to help men and women in the week. Let him not preach essays, or theories, or mere doctrine. Let him not fling the latest novel, poem, theory at ordinary people's heads. Does what he says touch their life? That is the test of his sermon. Would it be what Jesus would say to them if they took their troubles to Him?

Why, he was to ask twenty years later, should a London congregation in A.D. 1938 be enjoined not to covet its neighbour's ass?

He made it his business through fifty years of preaching to bring men and women into an immediate imaginative relationship with Christ. They were to be made to feel the impact of the presence upon the quotidian details of their lives.

The presence was to be felt, specifically. Les had no inhibitions about appealing to people's emotions: he believed religion consisted in large part in emotion. It is, of course, a dangerous game, however—men are ashamed of their feelings, Englishmen especially, with their stiff upper lips, the legacy of Victorian seriousness in general and Dr. Thomas Arnold, Headmaster of Rugby, in particular. Men would sometimes come up to Les after a sermon with tears in their eyes and begin to plead, "Here am I, a hard-bitten business man . . ." and he would later remark, "They're the softest kind of all!" But no criticism of a preacher is more biting than that he plays on people's feeling for effects, that he is a tear-jerker. Like any other successful preacher Les endured such criticism, but his attitude to the matter may be seen in a sermon in which he defends Billy Graham from the same charge. He had been to Harringay and heard Graham address 12,000 people: "I found no emotion present of a cheap, meretricious kind," he reports.

> There were no sentimental stories and no attempts to produce the easy tear. That being so, what is wrong with emotion? If Christianity is falling in love with Christ, has anyone ever fallen in love without emotion? Can we imagine somebody advising a young lover and saying, "I should not marry her if I were you, you evidently feel deeply about it." How could anyone come into contact with the living Christ, and feel both His forgiving love and His relentless challenge, without the very deepest emotion?

Les's preaching method and aim cannot be better summarised than by his own description of a dream—a day-dream, obviously—in which, he says, he saw Jesus in a number of familiar twentieth-century contexts. Too many people, he felt, think of Christ as ecclesiastical and respectable. "Many people adore a kind of stained-glass idea of Jesus. They see him standing on a spur of the Mount of Olives, say, with arms outstretched, calling them to come to Him. Few see Him going about with an immoral woman called Mary Magdalene . . . We have got to make Him incarnate again." Thus in the dream Les brings the person of Jesus into a business office, a household with mother and children, the rooms of a student and of a typist: Jesus sees their secret sins and all their evil. It is a conventional idea that God sees everything, and it has been used perennially to curb the vitality of rumbustious children. But in the dream, emphasis falls not on the weaknesses and sins of the characters but on their unadvertised virtues. (There was a remark of Chesterton's about Dickens that Les loved to quote: "He would walk into the foulest of thieves' kitchens and publicly accuse men of virtue.") Christ points to the secret generosities of spirit in people, and thus challenges them to do better all round. These imaginery conversations began as part of a sermon Les preached in India; subsequently they became part of his book, *The Transforming Friendship*.

A great deal of the power of his sermons came, I believe, from the rendering of the Christian religion as essentially an immediate and vital relationship with a very good man, a friendship, in effect. This, men and women could understand. Chapters from the history of the Jews, controversies over the Anabaptists or the heresies of Pelagius or discussions about discussions about God—these things touch only scholars, leaving most men uninterested, wondering how long the sermon will last, whether it will be raining when they get out, and what there will be for lunch. Les used to report another "dream" in which an Anglican clergyman approached Heaven's gate when he died, understanding that a knowledge of the Athanasian creed was the important qualification for entry. Peter cups his ear and says, "Huh?" and the minister proceeds to recite it: ". . . neither confounding the Persons nor dividing the Substance—the Glory equal, the Majesty co-eternal—the Father eternal: the Son eternal: the Holy Ghost eternal. And yet there are not three eternals, but one eternal. And in this Trinity none is afore or after another. But the whole three Persons are eternal and co-eternal . . ." At length in the dream Peter looked up, shook his head, and said, "You carry on. I'm going fishing!" Les had no great interest in church history or in the wranglings over complexities of doctrine which called, no doubt, for such carefully worded and qualified statements, skilfully contrived and patched by churchmen who were not mere pedants but wanted to get things settled in relative amity. But he used to imagine Peter and John sitting in the back of a church, nudging each other, whispering, "He wouldn't talk like that if he'd known Him!" And he used to pass up most of this kind of material, to speak of the human Jesus.

He was, of course, criticised, and sometimes severely, for changing or bypassing the traditional Christian beliefs and traditional interpretations of the Bible. His closest friend would concede that on occasion he was more than a little cavalier in sweeping away old ideas. Doctrines that had been deliberated upon and debated by generations of sages and scholars—honest men, men of solid learning, intelligence, and insight—he might flick away as if they were so much cobweb over a window. He questioned the truths people held dear and shook the settled beliefs with which they had made themselves at home, and naturally enough they came back at him. There was, for instance, quite a fuss over his treatment of the Atonement. Of a text in Hebrews, "Without shedding of blood is no remission of sin", Les had said, "In our modern view this is simply not true." A lot of people thought it *was* true, though; and, understandably enough, they also thought they were modern. People always do. In the newspaper in which the sermon on the subject had been printed, there was a steady storm of criticism. To one correspondent, the sentence from the Epistle conveyed "a profound and eternal truth"; another says, "If Mr. Weatherhead's assumptions are correct we Evangelists have been deluding ourselves . . ."; another, a little credulously, that she had seen "school girls, business girls and young boys whose very faces have shone with joy because they have believed that 'Christ's blood was shed for the remission of their sins.'" Another letter-writer says that, "in common with thousands of your readers", he is amazed at Les's theories and considers them to be "devastating to the faith of many . . ." Another, in "bewildering amazement", waxes quite rhetorical: "All is gone," he says. "All becomes an idle dream—the foolish imagining of simple souls. Was John Wesley deceived? Have our hymn-writers been deceived in their immortal songs? Was Saul of Tarsus deceived? Have we all been deceived? If so, then let us drape our pulpits in black and mourn over a lost Saviour."

Well, this was the kind of passion Les could stir up in people. Much opposition arose, no doubt, out of the fear that he was right and that the old order would have to change. But often the criticism was cool and intelligently and reasonably founded: Les's new ideas weren't always well-based, I daresay. He was not impervious to criticism; I think he was injured by hostility, as any sensible human being is. Sometimes, opening the mail at breakfast, he would be a bit disgruntled and snort at some criticism from the Church of England newspaper, the *Church Times.* Or it might be a half column of instruction from the *Protestant Vanguard* which under a headline "Insolent Blasphemy of a Methodist Minister!" once reminded him that he was "paid to preach the Gospel of the Lord, and not to ask silly questions or to make himself a stench in the nostrils of the followers of the Lord Jesus Christ". But no concern for criticism ever prevented him from speaking his mind, coming through loud and incisive, with no ambivalence. Sometimes I'm pretty sure he relished a fight, and after a speech or a sermon, he would say with some glee, "I shall get it in the neck for saying that!" I can see him now on a Sunday night, leaning forward in the passenger seat of the car, his black Anthony Eden hat tipped back off his forehead, cupping his hands round a match to light the cigarette which, after preaching, he used to say was the most satisfying smoke of the week; he would be perfectly relaxed and at peace with the world, and he would be expecting to get it in the neck. And when we got home my mother would appear and report that the newspapers had already been on the phone. Lapsing into some primitive dialect they sometimes shared for amusement, she would demand, "What 've you been 'n' gone 'n' done now?" pretending to scold, but smiling and proud of him. Les got it in the neck quite often: for his views on war and peace, on capitalism, contraception, fox-hunting, eugenics, euthanasia, and adultery, in addition to all the off-beat things he said about orthodox Christianity. People could get pretty tense with him, but he wasn't to be made to pipe down.

The message of the human Jesus he conveyed in simple language. As we have already seen he loved language. He admired its deft usage, *le mot juste*, the well-turned phrase. While he lived in Manchester, being in any case physically close to the university, he attended lectures and earned an M.A. in the study of English literature, particularly nineteenth-century poetry. But he had always been fond of the language of poetry (poetry, that is, that *he* considered poetry—"none of your modern stuff"), especially the rich diction of Tennyson, with its lovely suggestive rhythms and cadences. He quoted it often in sermons, and quite often he brought his own language to a poetic pitch. He might speak for example of the thrill of spiritual ecstasy—

> thrills kindled by some impassioned speech or mighty harmony, by the lisped prayer of a little child, by the sound of children's voices singing an evening hymn; kindled in some of us by the colour in the heart of a flower, the twitter of birds at dawn, snow-clad peaks aglow with the fire of sunset, the sound of the sea at night. And sometimes that experience is so deep and full that one feels that flesh and blood could stand no more.

Or there is this:

> Thus to love, thus to serve, thus to follow, thus to sail in unknown seas with the breath of life in one's nostrils, the wine of life in one's lips, the joy of life in one's heart; this is to find what life is, and why

> man was created in the morning of the world, when all the stars of God sang together; this is to find the joy Christ had—a joy unquenchable through all the sorrows of His earthly voyaging; this is to find what Paul meant by the "glorious liberty of the children of God". This is life. This is conquest. This is adventure. This is religion.

Les could talk and write like this sometimes. It may seem a bit psychedelic, but we have to remember that the purple passage was not always so unfashionable in writing and speaking as it is today, when our dearest literary efforts are as dry, terse, and flat as a stock-market report on a day of medium trading. The writers Les read as a boy and those who were considered exemplary in his youth were less inhibited than we are. Joseph Conrad, for instance, a sufficiently respectable model, can talk about the national spirit thus: "Like a subtle and mysterious elixir poured into the perishable clay of successive generations, it grows in truth, splendour, and potency with the march of ages. In its incorruptible flow all round the globe of the earth it preserves from the decay and forgetfulness of death the greatness of our great men, and amongst them the passionate and gentle greatness of Nelson." Well, Les ought to be permitted to be as enthusiastic about religion as Conrad about Nelson. He liked a bit of this sort of thing on occasion. But he was a more passionate man than most, to whom religion was more passionate than to most.

His great gift in language, however, lay in simplicity—the straightforward statement, honest and sometimes blunt. He was suspicious and contemptuous of highfalutin language. "What does he mean?" he would ask, and, on being answered, "Well, why doesn't he say so!" Or, "Is that the way he talks to his wife?" Throughout his life he had nothing but scorn for obscurity or nonsense in literature, in prayer, in hymns, or in the Bible. "I can't understand it," he would say aggressively of a poem of the early Eliot or Dylan Thomas, implying, "Nor can anyone else." He was really a bit mischievous in responding to literature, determined when it suited him to take figurative language literally and to ignore any compromise with meaning that literary form might require. All the same he had a number of points. He wrote to *The Times*: "Dear Sir, Since the hymn, 'O God, our help in ages past,' was recently mentioned in your paper, may I plead that if it *must* be sung on national occasions, two verses at least be omitted? One says, 'Sufficient is Thine arm alone,/ And our defence is sure.' The fact that we do not really believe this is evidenced by the representatives of the armed forces present on all national occasions." The other verse referred to was that in which time, like a rolling stream, "bears all its sons away", which Les found incongruous for a day celebrating *remembrance*. He also used to snipe occasionally at the Prayer Book:

> How can anyone pray [he wanted to be told] "O Thou Who alone workest great marvels, send down upon all bishops and curates, etc." as though for bishops and curates to be helped would indeed be a marvel? How can anyone pray, "Give peace in our time, O Lord, because there is none other that fighteth for us, but only Thou, O God"? In other words, we must have peace whatever happens because our only ally is God, as though He were indeed of very little use.

At the marriage service, he noted, "the married couple are told that marriage exists so that a certain sin will not be committed, only to find that when they bring their baby to be christened, they are told it was conceived in sin after all."

In a letter to the *Daily Telegraph*, Les quoted Psalm 68, saying that if it represented the mind of God he would never worship him again: "God shall smite through the head of His enemies, the hairy scalp of such an one as goeth on still in his guiltiness. The Lord said, I will bring them again from the depth of the sea: that thou mayest dip thy foot in blood; that the tongue of thy dogs may have its portion from thine enemies." People wouldn't come to church to listen to such drivel, said Les, and who could blame them?

He used straightforward language in the pulpit, as simple as his paradoxical and mysterious subject permitted: as he used to remark, "Jesus stood up in a boat and talked about His Father." His talk was personal. He gave the impression in the pulpit that he was talking to each individual personally; frequently a correspondent would confess, "I felt your message was for *me*." "He just leans down over the pulpit and talks to you," one old woman said of Les. She was the charwoman who used to come and "do" for us at the Dial House, the holiday place we had on the Kent coast. I think a lot of people felt the personal touch and the simplicity of the message and knew it as the essence of Christianity. He always recognized there were paradoxes and mysteries beyond human comprehension—where they had to remain. He could only dispose men's souls toward God, spiritual health, and the good life.

Along with the plain language went the simple illustration—the illumination of a concept or an attitude by something very familiar in ordinary life. The word "gospel" he said, meant "good news" and good news was that which could be shouted across a street; ". . . the war's over! The baby's born! John's got his matric! Susan's out of danger! The strike's settled!" Then, leaning down over the pulpit, "Here is my bit of good news for you: God will receive you—Now!" Sometimes he would take examples from natural history, an anecdote about birds, for example, in the behaviour of which he was particularly interested, or an odd bit of lore like the following, with which he opened a sermon at the City Temple:

> There is a curious creature called the caddis worm. One of his most curious habits is that wherever he wriggles he picks up little bits of metal or stone or grit and puts them on his body to protect him against the jolts of his life and the bumps of his journeyings. One of the great dangers of modern religion is that we should use it for that purpose only . . .

Or there was his own favourite parable of the Persian carpet, the kind that had arrived on the study floor by the sequence of events already described. It is made on a vertical frame, the artist standing one side and the boys who do the weaving on the other. When a boy makes a mistake, the artist does not necessarily have him change it; he alters the pattern. "I wondered," Les used to say, "if that was not a kind of parable of life." We are the boys working on the blind side of the rug; we make mistakes; but in the end, when we see the finished product, we will understand that no mistake of our has been allowed to ruin the overall effect.

Or there was this:

> Let me illustrate this extreme sensitiveness in the mind of a child by reminding you of your own experience when you were first given a box camera, with a film in it, and you tried to take a photograph. Armed with the camera you went out into the back-yard and saw the cat sitting on the dustbin. You pushed the lever and took a photograph of the cat. Then, without remembering to turn round the film, you took a photograph of your aunt, and afterwards, when the film was developed, you could not tell where the cat ended and the aunt began. I think I am right in saying that there is no process known to photography by which the aunt can subsequently be entirely separated from the cat.

With the aid of this illustration, if you had a cat, a dustbin, and an aunt, you could easily imagine the sensitiveness of a child's mind!

It is not easy to limit these examples of Les's art of illustration with which he fleshed out the dry ribs of his metaphysic. Others will appear in these pages, but the following ought to be cited:

> I want you now to make a mental picture of your fishmonger, the man who last Friday wrapped up some cod in a newspaper and handed it to you! Don't think I am criticising fishmongers. Ours is a friend. But if on the way home somebody said to you, "Do you know that the man who wrapped up your fish today will one day be known by his Christian name to every educated person in the world? The most influential cathedral in the world will be named after him and he will write words that will be read in every church for two thousand years. In some people's minds he will not only be called a saint, but be regarded with the deepest reverence. People will not only revere his statue, but kiss its toe," would you not laugh and say, "Don't be ridiculous"? But all this is true of Peter who wrapped up in leaves parcels of fish and sold them to men and women in Galilee. And what was the transforming power that changed Peter, the fisherman, into Peter the most famous apostle in the world? The answer is the dynamic personality of Christ.

A young girl being examined in theology preparatory to a career as a missionary, having been instructed to compile a list of relevant books she had read, put Les's books under the heading "Fiction". "They're so easy to read," she explained, "you could hardly call them religion."

During the Manchester years, he began to bring his knowledge of psychology into his sermons. His interest in this discipline had begun in the desert, when a doctor practising psychotherapy in a hospital ward had turned to him and said that most of that kind of work ought to be done by the padres in the regiment. He was a man of brilliant and magnetic presence, who had cured a soldier who, through shell-shock, had lost both speech and memory. Les was made to understand that the man's dumbness was due to his fear of admitting his own fear. The doctor related these psychosomatic troubles to those in the gospels recorded as cured by Christ. Les listened to this man expounding his theories, as they would sit together talking deep into the night. Often, after such a session, he couldn't sleep for excitement, and he would lie awake watching the yellow glow of the lamp through the mosquito net, imagining the wonders of this new territory where science and spiritual exercises came together. A few weeks after Les's introduction to him, the doctor was thrown from his horse; his head struck a tent peg, and he was killed outright and buried at sunset the same day. Thinking of the theories and the secrets the doctor had imparted, Les felt the responsibility of a great inheritance.

Back in India he began to read Freud, Adler and Jung, Ernest Jones and Rivers, and he discussed psychological healing with physicians and psychologists. He became increasingly familiar with the subject: he was one of the first to relate psychology with religion and must be given credit for having helped to disseminate facts and ideas of what was then a new science.

His use of psychological methods for healing will be discussed later. For now, we may see how he used psychology regularly in sermons to explain some of the activities of Christ in the gospels, or simply to demonstrate the curious workings of the human mind. He believed firmly that God guided men in dreams: psychological research was thoroughly in accord with the lines from Job which read, "In a dream, in a vision of the night, when deep sleep falleth upon men, in slumberings upon the bed, then He openeth the ears of men, and sealeth their instruction." Les occasionally gave examples of dreams that had been related to him which, when decoded, seemed to contain spiritual guidance. A young minister, for instance, dreamed that he confidently mounted the pulpit steps to preach, only to find himself wearing a pair of scissors instead of the cross. In the gallery on his right he sees his mother, and she is looking uncommonly old. Interpreted, the dream is revealed as a warning not to engage in a questionable and divisive financial deal, symbolised by the scissors, which would disgrace him in the eyes of his mother, always in the right, and bring about her premature ageing.

Sometimes Les used psychology to explain biblical events, or at least to make interesting guesses at their meanings. The lunatic in the story of the Gadarene swine declared that his name was "Legion". Les thought it possible that at some time or another, perhaps in childhood, the patient had suffered a shock at the hands of the Roman Legion. Other instances could be cited to show that when a person has been driven mad by a shock, he will repeat some word connected with it. During the war a man was found wandering in No Man's Land who had been tortured by the Germans and could only repeat the word "Boche", slang for German. Possibly the man in the New Testament had witnessed the Massacre of the Innocents: Les imagined him as a child seeing the Legion marching through the streets with drawn stained swords, and sustaining a shock that might have been sufficient to induce madness. The word "Legion" might then well be the word which expressed, in summarised form, the origin of the insanity.

Finally, a good deal of the delight in listening to Les's sermons was due to his sense of humour. He possessed a luminous wit and a sense of the grotesque; he loved fun. A colleague once applied to him terms used of Gladstone—"a wonderful man and full of boy". He had found plenty of evidence in the gospels that Christ himself had an abundant sense of humour, imagining Pharisees straining a gnat out of the soup and swallowing a camel, a man declining a supper invitation because he had to try out some oxen—*at night*; a man lighting a lamp and shoving it under the bed; a man with a two-by-four sticking out of his eye groping for a speck in another's. Les used to say that humour was a divine gift and could be beneficial to human kind. Laughter could be a spiritual activity: "It makes *Punch* quite as spiritual a magazine," he once wrote, "as the *Journal of the Wesley Bible Union*, though not always as funny. *Punch* makes one more in love with life." He would rather, he says, see Charlie Chaplin than hear "a *thoroughly* doleful *preacher*". He had a strong sense of the burdens that people have to bear throughout their lives, and he felt that humour was of spiritual, not merely secular, assistance. During the war, particularly, when burdens were universally heavier, I can remember him leaning over the pulpit and saying with some urgency, "If you know a funny story, for God's sake go and tell it to someone."

Naturally he felt no compunction about the employment of wit in the pulpit. But no: "Employment" suggests something more calculated than his practice: he didn't plan to be witty; wit came spontaneously, irrepressible and unprepared, surprising his congregation and sometimes shocking it, shocking quite frequently my mother who would expostulate a little afterwards and purse her lips and say, "Oh, Les, you *were* awful." Les was apt to gesture quite a bit in his sermons: the *one* God, the *one* great commandment, or the one anything else might be accompanied by the arm being raised vertically with the forefinger pointing up; a challenge might be accompanied by a finger pointing into the congregation at some arbitrarily selected member; for large conceptions, such as the glory of the Kingdom of Heaven or the fall of the Roman Empire, he would stretch his arms wide apart and let them fall slowly to his sides, trailing the sleeves of his gown. Once when he was preaching the university sermon in St. Giles' Cathedral, Edinburgh, all the lights had been switched off except one which burned above his head. Suddenly, with one outflung arm he contrived to switch this one off too, plunging the cathedral into darkness. For a moment the blackout produced a breathless silence in the old church until he remarked casually, "I seem to have extinguished my halo," which brought a gust of laughter from the student audience, and later, in the vestry, brought from one of the elders the observation that no one had ever laughed before in St. Giles!

Out of the pulpit, Les was regularly full of wit and fun and tomfoolery. He had read *Three Men in a Boat*, by Jerome K. Jerome, and for fifty years he thought it the funniest thing in the world. It describes a holiday in which three men and a dog go boating on the Thames. They run into hilarious and preposterous situations and solve them or fail to do so by hilarious and preposterous methods. Most of the book is written in a dead-pan style, and of the three characters the one with the most dead-pan manner is Harris, and it is Harris who has the driest of the three dry senses of humour; Harris visits graveyards in the towns "to enjoy the tombs"; he is the master of understatement; and when any one of the others makes a joke he remarks laconically that it is old. He appealed to my father immensely; their wit ran along the same kind of lines. It was not the sort of wit, however, that everybody instinctively finds amusing: it didn't always occur to Les that some people enjoyed other and less mordant varieties or that people might have lived reasonably satisfactory lives, in no sense intellectually deprived, without ever having read *Three Men in a Boat*. He would allude to it as if it were stock reading and as familiar as the twenty-third psalm or Hamlet's speech on suicide. If, on hearing someone else's joke, Les should remark, "Harris said it was old," he might be a little surprised at the manifest perplexity he encountered, although the world is full of well-meaning people who have never heard of Harris. Once, as we were all in the car driving into Dover, he saw the daughter of our village grocer emerging from the cemetery and he hailed her with the query, "Have you been enjoying the tombs?" assuming she would relish a reminiscence of the old hero. But the allusion to the cultural heritage was lost on her: she looked a bit puzzled and said she had recently buried her father. Les was pretty crestfallen at this.

In the novel, on an occasion when one of the characters is conveying some cheeses by train to London, a man in the compartment says there is a curious smell of dead baby. Les once tried this one out too, but it went over like a lead balloon—that audience hadn't read the book either! Another work that amused him intensely was *Balbus the Bat*. Some things that I thought were amusing, on the other hand, left him cold. I gave him *Lucky Jim* by Kingsley Amis, which was the funniest book I had read for ten years, and he dismissed it, saying he had "conceded" it a grin at one point! When I sent him *Gamesmanship*, he wrote as follows: "Many thanks for the Gamesmanship Book. I've read 35 pages and now find that it was meant to be funny. They ought to tell you these things on the cover. I thought it was meant to be sad, and, as I am easily moved to tears, I wondered at my own fortitude in resisting even a sniffle let alone a sob. Now I know what the book is really *for*, I think I shall have many an uproarious half hour alone in the study with it with a wild wind round the house and the rain lashing the windows. Thank you!"

He made it a point to tease people as soon as he knew them well enough, or, what was sometimes a little earlier, as soon as he *thought* he knew them well enough. We occasionally lost a gardener or had to change plumbers on account of their singular inability to appreciate good clean intellectual fun. But you really had to know him pretty well. Often his remarks had to be edited. His secretary, tactfully putting her hand over the telephone, would say, "Mr. So and So wants to speak to you." "Oh, tell him to go to blazes, " Les might say; and she, infinitely trustworthy, would forthwith translate his response into the more conventional terms of social intercourse. Once in a sweaty summer, as he was about to fly back from New York, he was asked by reporters for a comment on their city and he said that he thought it was pretty much like Hell. He didn't mean any harm—just a sort of joke about the weather; but before he had unpacked his brief-case in London, editors were on the trans-Atlantic telephone trying to get from him a considered statement about the morals of New York and a withdrawal of his awful denunciation. You really had to know him.

His abiding friends were not thick-skinned but they had learned to put up with his raillery—they learned, I guess, to see it as part of his charm. Mrs. Keith Robbins, who was the same age as Les, was an old friend, and once in Bexhill, after a lunch party, he said to her, "Keith, you're going to remember me in your will, aren't you?" My daughter, then a young child and moderately well brought up, was present, and her eyes grew large in amazement. Keith replied, "You know I am, Leslie." "Good," said Les, and then, to the child's increasing horror, "Well, leave me some time to enjoy it, won't you?"

There was an inexhaustible supply of mischief and fun in him, and his family derived endless amusement from it. They derived also a certain amount of embarrassment, of course, especially from his whisper. This was invariably a stage whisper, perhaps because he simply miscalculated the number of decibels necessary for communication, but also perhaps because he was used to addressing a thousand people or more and wasn't going to waste a witticism on a limited audience. If we stayed in a hotel, he must needs bestow nicknames on all the other guests, and no one except Les could possibly believe that they didn't all learn them before we left. Often in public places he would make loud laconic remarks. Once in a railway compartment we were joined by a small bald gentleman who immediately upon entry opened the window. Les had pretty firm views about open windows: they should be closed. And feeling no doubt that since we had arrived first in the compartment it was we who should determine its degree of ventilation, he promptly stood up and closed the window. Shortly, the newcomer opened it again, to a slightly less degree, it must be said. Les closed it with a bit of a slam this time, and announced to me in the usual whisper, "If that bald-headed little coot opens that window again, I'll shove him through it." He surely didn't want the little man actually to hear him; but you could say that he wouldn't mind if he should catch the general drift. Over their fifty years of marriage, my mother must have spent many anxious minutes vainly hoping that Les wouldn't say anything in one situation or another and then, when he did, vainly hoping that he hadn't been heard. "Les, you *were* awful," she would say later. The rest of us found his behaviour highly entertaining and were, no doubt, accessory to his crimes.

Once when he was old we were walking along the promenade at Bexhill, a town of old people and old dogs and cats, when we came up behind the most singular-looking animal I've ever seen, being towed on a leash by an old lady. Les pointed at it with his walking-stick. "Oh dear," he said. "Oh dear!" I knew some final judgment on the poor ill-favoured beast was in gestation, and I tried to lead him over towards the sea where a chap in a sailing boat was attempting to shake out a spinnaker. But it came, audible up and down the coast, the familiar stage whisper: "A brute as ugly as that, even if you loved it, you'd only take it out at night."

Well, he could be like that. He could be solemn when occasion demanded, as it often did; but he loved fun, and it would come bursting out. It was provoked especially by false solemnity, which was a red rag to him or a bubble he had to prick. "God is not a bit impressed," he wrote, "that you've got 'Hon.' in front of your name, or M.A. after it, or 'Fitz' in the middle of it. Come off it!" And once in the vestry before church, doing valet service to one of his most distinguished frock-coated church officers, a man of aplomb with solid investments in the City, he said, "Stick that out and that in," tapping him on the chest and on the rear with the clothes brush. "And when your mother has got you ready for church, don't drop crumbs on your waistcoat." On one occasion he was delegated by the *British Weekly* to report the proceedings of an Oxford Group house party. The major feature proved to be a game in which, the sexes having been segregated, each person in turn confessed his major sin—a good clean moral exercise if ever there were one. Unfortunately, however, these people, so serious about their goodness and stiff in their dinner jackets and piety, triggered off the old iconoclasm in Les, and he wrote that he had found the sessions valuable because by the time it got round to his turn to bare his soul he had learned two new actual sins and a handful of temptations. Frank Buchman never forgave him for this.

Once in a meeting, sitting on the platform alone with Lord Soper, he was the victim of one of those endless extempore prayers offered by a layman. The brother was ranging from pole to pole requesting blessings for the whole human race, listing the poor, sick, hungry, and stricken of each state and the endemic distresses of the various ethnic groups, introducing each brand of disaster with the phrase "Thou knowest O Lord", disarmingly suggesting that the whole output was supererogatory —Les scribbled on his hymn sheet, "Will this blighter ever shut up?" and passed it to Donald Soper, who fumbled it, letting it flutter to the feet of the cosmic intercessor himself, who then picked it up and passed it on. Whether he read it or not they never knew. But the chances are he did; he shut up quite soon.

Les did things like this. For great men and women—not necessarily great in the world but people of genuine intellectual ability or moral courage or kindness—he had unbounded respect. But for false solemnity, self-esteem, pomposity, fake piety, and anything else that smacked of the phoney, he was the iconoclast: for these there was always the schoolboy lurking in the wings, thumbing his nose at the Lord Mayor's procession, ready to put the tack on the seat of the fat committee chairman or audibly sniggering when wealth, pride, or portentousness went down to the banana skin. Not everybody saw this side of Les, and it must be painted in with some bold strokes. Thousands of people have seen that solemn figure slowly climbing into the pulpit, head bowed in the almost insupportable sense of his own unworthiness to speak, and to speak to so many of the awful truths of the God whom he revered and loved. But far fewer have seen the other, the complementary side, which does not contradict but makes the picture whole. This side of him, the wit and the humour, the buffoonery and the mischief and general rumbustiousness, was in a way an index of his profound seriousness and not a qualification of it. He was the lovable clown sometimes *because* at others he was the humble, devoted and holy man. There were places, vast places, in life for both. That was the way Les saw it, I'm sure.

There are two more things, however, that must be suggested as related to his power as a preacher. One is the quality of his belief in Christianity and its Founder. His belief was strong, like that of other members of the profession, but it differed, I think, in vividness. For some ministers, Christianity provides a carapace of principles beneath which virtues that are not necessarily fugitive and cloistered may be nourished; for many, belief provides a system of morals, a code of ethical conduct, a way of life to be followed; and their work in alleviating grief, comforting the distressed, proclaiming what is right and wrong and in general injecting an infusion of decency, unselfishness, and visionary living into a world given over to greed, violence, and fear—their work makes an immeasurable contribution. For Les, there was the Augustinian injunction: "Love God, and do what you like"; and moreover there was the living Christ, who, if he was the source of moral principle and ethical behaviour, was equally the sun-tanned figure who roamed around Galilee, who lived in history, and lived now in Les's imagination. And the influence of Christ was not only intellectual, a conviction in the mind; it was also felt powerfully in the blood. Here and there in his writings he has described occasions when a sense of the presence of Christ flooded his consciousness—once, for instance, coming from Oxford at night, bringing him to stop the car and get out and stand among the dark trees by the road to worship; and once, as described later in these pages, in a train at Vauxhall Bridge station, a singularly unconventional background for a visionary experience.

He believed powerfully and vividly. He had doubts, sometimes, of course: they assailed him especially when his physical health was below par or when he was depressed mentally. Thus in 1951, writing to me from Australia, he says he often gets downhearted. "Where, oh *where* is the secret we are all looking for? Is it in Xy. [Christianity] and we have missed it or is it partly in those strange eastern systems like Yoga? My host here [Raynor Johnson] is very keen on the mystics and Yoga and psychic research, rather interesting in a London D.Sc. in Physics. I wish I knew the answer or is it that what one needs is not to *know* more but guts to *do*, up to the point of knowing." The letter goes on to describe how he had just received a public welcome in Melbourne from 3,000 people: "They wouldn't stop clapping," he reports. It must have been some compensation.

The same self-doubts appear in a letter to James Stewart in 1957. Les was in some despair after the repetition of an operation still seemed to bring no healing. "I have read a lot about healing," he writes, "but . . . it doesn't seem to *work* in my case." He thought his physical ill-health might be attributable to his own spiritual failings.

Certainly there were dark times interspersing days of clear beliefs. One of his favourite quotations was that bit in Browning—

One who never turned his back but marched breast forward,
  Never doubted clouds would break,
Never dreamed, though right were worsted, wrong would triumph,
Held we fall to rise, are baffled to fight better,
  Sleep to wake.

He liked these lines and mostly he lived up to their example. But he wasn't always quite sure of the ultimate benevolence of the universe, and sometimes, to express a different mood, he would rather quote Walter de la Mare: "'Is there anybody there?' said the traveller,/Knocking on the moonlit door." Doubt, in some sense, is an index of faith, and the saints have nursed it in the dark night of the soul. Les had absolute faith in life after death: he never spoke of death except as a new kind of experience of living with freedom from the body. At Christmas in 1972, he expressed his sense of the mystery of our human condition in a letter, saying, "I shall have many questions to ask God when I do reach the other shore, which I hope will be soon." Then: "L.D.W. is very well thankyou but the house he lives in is falling into decay." There were doubts, and there were visionary moments when he felt extraordinary manifestations of the presence of Christ. But most of the time there was the steady quiet conviction of the presence of God, like that of a human friend, to be called upon.

The other matter to be mentioned in accounting for his success as a preacher is his love of people and in turn their response to it. When later he became the President of the Conference, William Sangster said of him, ". . . if I were driven to put the secret of his power into one word, I should use the word 'caring'. He cares for people: honestly, personally, and in a most costly way. He cares." Sangster goes on to describe how Les's feeling for others came through when he spoke of people who were sick. "There is nothing mysterious about this man's power," he goes on. "People love him because he loves them. Even those who do not know him personally feel, as he preaches, an almost personal love reaching them . . . Frankly, I have often wondered how one human heart could feel so much." Once two other ministers were discussing the same subject, and one concluded thus: "I know his secret. He opens his arms to describe God's love and no one feels left out." It is one of the choicest comments on his life and work.

He made a great many friends through an organisation called the Fellowship of the Kingdom. He had heard about it from Newton Flew in India, and on his return he became one of the early members. It met regularly—a group of ministers in each district throughout the country, prepared to discuss a theme related to Christianity. Les loved these meetings; often they were held in his study, the men bringing sandwiches and making a day of it. It was good to spend the best part of a day in religious talk and informal chat with men he was fond of. Then, each year, Les used to throw his tennis things into the car and drive up to the general meeting in Swanwick, in Derbyshire, for a week of discussion and fellowship and so forth. He had good times at these conventions and used to return exhilarated, and one devoted member of the organisation recalls that "for years the chief attraction of the annual gathering at Swanwick was a chance for men to see Weatherhead and to talk to him".

The psychic mechanisms of the world are mysterious enough but it surely is not nothing to be loved as Les was, and indeed to be prayed for. We shall see below what extraordinary effects seem to have been brought about for sick people by the prayers offered in the church. But we may note here that Les was quite convinced that his own powers derived in no small measure from the intercession of others. "Again and again," he writes, "my friends have expressed astonishment at the amount of work I have been allowed to get through. When the books are opened I think the secret will be revealed. Greater than any mere doing, the purest of all activity, as Von Hugel called it, is the praying of lowly-minded men and women."

While he was in Manchester, Les submitted a manuscript to the publishers. It had been written in India when he had been moved by the terrible bereavement of men and women during the World War. It was called *After Death*. The publisher agreed to produce it if Les would put up thirty pounds to cover costs of publication in case of its failure. Les declined the gambit: thirty pounds was almost six weeks' salary! But my mother insisted that he should go ahead, though heaven knows how she managed to cut her budget, carefully maintained in the old red school-exercise book where she kept account of every household penny.

After Death appeared, dedicated to Leslie's mother and headed by a verse found copied in the fly-leaf of her Bible after she had died:

> If I should die, and leave you here awhile . . .
> For my sake, turn again to life, and smile,
> Nerving thy heart and trembling hand to do
> Something to comfort weaker hearts than thine.

The book discusses such questions as would have special relevance to those who had lost sons, husbands and fathers, presenting these matters as they must have been desperately hoped for: death as the entry into a fuller life, the condition of life after death, recognition and reunion, the growth of the soul. All in all, it is a liberal exegesis: hell is relative, temporary, and subjective; judgment is self-judgment; the forgiveness of God is absolute. In some respects this first book is characteristic of a great deal of Les's later work. It relies much, for its argument, on analogy. Most often that of the father comes into play: God as father must be conceived of as better in every respect than a human father can possibly be. "If ye then, being evil, know how to give good gifts unto your children, how much more shall your heavenly Father give the Holy Spirit to them that ask him?" Thus in connection with the idea of unending torture in hell, Les writes: "If a child were rebellious and wicked for ten minutes and a father tormented and thrashed, tortured and starved her for the rest of her life, his action would be merciful compared with that of a God, who for the sins committed in one lifetime tormented a soul unendingly." Often enough his position is rationally deduced from the assumption that God and the universe are rational and good. Thus he may write, "We speak of this consummation because it seems to be the logical outcome of any intelligent thinking about God . . ." The call upon logic and intelligent thinking is characteristic.

Occasionally he explicates a scriptural phrase by going back from the English to the original Greek, as we have seen him doing in his sermons, finding the special nuances of the word that have been obscured by translation. And occasionally he brings psychological insights into play, suggesting, for example, that in after-life, the memory of sin may be total, citing the recall power of minds under hypnosis. The argument of the book is embellished repeatedly by lines from the nineteenth-century poets, Longfellow, Whittier and, most frequently, Tennyson. (One reviewer suggested that Les might have consulted Ambrose and Origen rather than so many poets.) The style is beautifully clear throughout, ascending from time to time to passages of poetic expression and fine writing.

The issues he deals with in this book were very much in men's minds in those days, but it is questionable whether it would have had the publicity it did in fact enjoy if it had not been for the passionate hostility its ideas drew down upon it from a colleague in the church. The sub-title of the book is "A Popular Statement of the Modern Christian View of Life Beyond the Grave", but one of Leslie's fellow Wesleyan ministers declared that the theories were at variance with Wesley's interpretations of the scriptures. Speaking of the matter many years later, Les said, "The brother's dead now, so he *knows* I'm right!" But at the time he must have been a bit disturbed to learn that the question of his heresy was on the agenda of the discipline committee of the Wesleyan Conference. Before the Conference convened he was given a list of the charges, and he prepared his defence. But when he appeared before the committee, the charges produced were different from those which had previously been made. His objection was properly sustained, and the hearing was postponed for twelve months. In the end he was exonerated: at the next session of the Conference, it was decided in committee and ratified by the whole body of ministers attending that his rendering of the "Modern Christian View" did not go beyond the freedom of exegesis that the church allowed its ministers.

In the meantime, however, in the year between the original charges and the judgment, interest was aroused, and the book began to sell. It was readable; it was comforting; and it was under attack. It went on selling for years afterwards, though. In 1926, an anonymous churchgoer who admired the work gave Les two hundred pounds to purchase eight hundred copies to be presented to people in the church.

4

Les had been in Manchester only six months when an invitation came to him for September 1925, two years ahead, to follow the Reverend A. E. Whitham at Brunswick, a large Wesleyan church in Leeds. It was a massive square chapel, heavy, and, like most buildings in Leeds, black, surrounded by thick iron railings. Inside, a gallery ran round all sides; downstairs, some of the pews towards the back had neck-high divisions and doors, like little loose-boxes. The spacious pulpit, in splendid shining mahogany, dominated the whole of one end of the church, which was so easy to speak in that it was sometimes referred to as a "whispering gallery". Les's induction coincided with the centenary of the building, and he attended the celebrations, taking a small part.

It was no minor matter to follow a preacher as successful as Whitham. He had done well, and it was natural enough that a number of the church members and officials should expect a falling off. Les himself had no sanguine expectations of keeping up his predecessor's standards. But there was no falling off. After the first six weeks, the chief steward approached him in the vestry after the service and addressed him in the rich broad Yorkshire dialect: "Yoong man," he said, "you're doin' vurry well, vurry vurry well." He nodded approvingly, and Les thanked him for his words. Then the steward whispered, "Boot can yer keep it oop?" To which, with a perfectly straight face, Les replied, "No, I'm afraid not. I had six good sermons when I came here, and I've just preached the last." "Aye," said the other, "I were afraid o' that." That was Arthur Barrett, a solid pillar of the church and its treasurer, who owned a wholesale grocery business in the city. His spirits, Les used to say, went up and down with the price of lard. On Sundays he kept an eagle eye on the collection plate. "There were a woman in church this morning," he once reported, "who 'ad on a fur cooat—a fur cooat, mahndyer. And what d'you think she put in t' collection plate? A penny!" All the same, the collections continued to be substantial. One Sunday after the money had been counted, Arthur Barrett came into the vestry, again bearing the ledger in which the takings were entered from Sunday to Sunday, month to month, and year to year. He opened it at two places, bending back the pages in between to compare the recent collections with those of the same month during Whitham's last year. "Look," he said, pointing to the two sets of figures, nudging Les with his elbow and spilling the secret like a conspirator. "You beat 'im there"; then, moving his fingers down the pages, "You beat 'im there; and you beat 'im there. 'E beat you there, but you 'ad it wet, and they doan't coom aht in t' rain!"

Les had asked the church if they would provide him with a motor cycle for visiting the sick. But in the stewards' meeting, Arthur Barrett declared, "We doan't want our minister visitin' ladies' bedrooms wi' grease all oover 'is 'ands and clooathes. Let's get 'im a little caar." And so a little car it was, Les's first, a bull-nosed Morris with a round brass radiator, a bulb horn outside the window, and a dickey-seat at back. After that, every New Year's day, Les would go down to Appleyard's, the Morris distributor in Leeds, to turn in the old car and bring home a new one, accompanied by his two small boys, who fingered the chrome and revelled in the smell of new leather and the incredible shine in the coachwork. As far back as I can remember, our cars in Leeds were green, dark green, and they were washed every Saturday morning for half-a-crown by a chap called Horace, who had been turned out of a church choir for something or other and whose sister had thereupon called upon the minister and declared she was an atheist!

Les was thirty-two when he began his ministry in Leeds and he directed it to a large extent to young men and women. He dwelt repeatedly on the potential of the young, their opportunities and the responsibilities they would face.

He used to use the word "radiant" to describe the young—the word which so often designated the true Christian way—and he associated them with health and activity in the sun. He approved of their hiking on Sunday (not of their playing games, however) and their sunbathing. At the end of the long hot summer of 1933, after there had been widespread protests from the watering-places of England at the decreasing extent to which bathing suits concealed the skins of bathing people, Les responded in the press with a spirited defence: women wearing low bathing suits that revealed lovely shoulders were not therefore immodest: God had fashioned the human body, and there was nothing intrinsically immoral about it; the rigours of the climate, without any appeal to Mrs. Grundy, would automatically preclude complete nudity, he thought (never dreaming how warm things were going to get in the seventies); the young were not to be inhibited by the old in these matters. Often Les saw things in terms of the familiar dichotomy: youth, the victim of wicked age, like Christ Himself who was young and had been done down by old respectability and aged custom.

In 1929 he accompanied the Bishop of Ripon to Germany as a part of the World Alliance for the Promotion of International Friendship through the churches, and he was impressed with the young *Wandervogeln*—the German youth, wandering through the forest in bands, staying overnight at the *jugendherberger* hostels, swimming in the rivers, absorbing the sun, and in the evening re-creating the old folk songs and discussing among themselves the folly of the war and of the old men who had brought it about. It is ironic to recall that ten years later he was organising his own peace movement with Germany, sending over members of the City Temple in what he called a Friendship Invasion in the summer of 1939, until, late in August, war being imminent, the Foreign Office wired him to recall his people, the telegram arriving in the evening in the last slanting rays of the sunlight on the garden. And it is ironic to think that the youth he so admired in the late twenties would, in the early forties, be the same men up there in those throbbing Dorniers that bombed his church and, night after night, his city. But then war breeds every kind of irony—for another example, this: Martin Gerhardt, a young minister he met in his 1929 visit to Germany, was exactly the same age, had the same number of children, and was interested in the bearing of psychology on religious work; they were united by a score of other common interests. But during the First War Gerhardt had been in the same theatre as Les, leading a company of Turks as Les had led his Indians, and if they had encountered each other then they would have done so with rifle fire.

Les wished very strongly to make his church attractive to the young. And he wanted the attraction to be of the right kind; not billiards and tennis-clubs, but a challenge like Garibaldi's—"I offer you hunger, forced marches, wounds, death." Most of all he looked to the young to prevent another war. Again and again he called upon them, the "young men of all nations with the brightness of the morning in their hearts", to band together for peace and reject the warmongering machinations of the proverbially wicked old men, the politicians, "the cunning schemers for world power whose soul the devil had bought"—it was quite the vogue in the twenties, as at any time since, for the young to brand the politicians in this fashion. And it is amusing to note that one young man who attended a service at Brunswick and was able forty years later to recollect the exact date, the text, and the substance of the sermon, was himself destined to become a politician of some note in England—Harold Wilson.

A good part of my father's message in Leeds, especially in the darkening days of the middle thirties, was pacifism. Throughout the country indeed, in the late twenties and in the thirties, a widespread sentiment repudiating

The City Temple after enemy action in 1941.

Despite the bombing the City Temple goes forward into its fourth century.

war was expressed by Dick Sheppard, canon of St. Martin's in the Fields, and his Peace Pledge Union; by George Lansbury, leader of the Labour Party, and the leftist intellectuals; by Aldous Huxley, and Professor C. E. M. Joad. The Oxford Union, under the influence of Joad, passed a resolution that "This house would decline to fight for king and country." The feeling was so strong that the Prime Minister, Stanley Baldwin, considered it necessary, though against his better judgment, to forswear re-arming in order to win the 1935 General Election.

Les was concerned that a young generation was emerging from adolescence that had not known the horrors of war but looked upon it only as a field where gratifying emotions might be stirred and glory won. In February 1932, coinciding with the first day of the Disarmament Conference in Geneva, a massive demonstration was held in the Royal Albert Hall in which the different denominations of the Church were represented. A long procession moved down through Whitehall and Westminster and a huge gathering mustered in the hall itself. It is curious to see in the press photographs Les's young pale features beside the plump square head of William Temple, then Archbishop of York, looking a bit like Teddy Roosevelt in his earlier years, and the distinguished figure, domed and white-haired, of Cosmo Gordon Lang, looking every inch the Archbishop of Canterbury. In his address, Les related the anecdote about the man in Leeds in the First World War who got three months' hard labour for distributing 20,000 copies of a leaflet which, said to be seditious, was found upon inspection to contain nothing but the words of Christ, taken without comment from the Sermon on the Mount. He addressed himself particularly, however, to the young, anxious to have them recognise the reality of war: "You have never seen a bayonet driven into an abdomen and twisted round and pulled out," he said. "You have never seen boys of nineteen blown to pieces and threads of their bodies hanging on barbed wire. And not only are men involved, but also women and girls go into factories to make explosive shells which shatter the bodies of other women's husbands and sons and sweethearts into unrecognisable pulp."

Malcolm Muggeridge attributed the pacifist movement in the thirties to the fact that modern war endangered policy-makers as well as the military. "Kipling's popularity," he wrote, "has waned as air-raids have become more dangerous . . ." This may or may not be so. But there is no doubt at any rate that pacifist journalists, politicians, ministers and other spokesmen made much of the anticipated horrors which the next war would produce in the way of germ warfare, gas, and aerial bombardment. Churchill's rhetoric resounded in quotation here and there:

"Mankind . . . has got into its hands for the first time the tools by which it can unfailingly accomplish its own extermination. Death stands at attention, obedient, expectant, ready to serve, ready to shear away the peoples *en masse*; ready, if called on, to pulverise, without hope of repair, what is left of civilisation."

Again and again Les called passionately upon the young, setting their brilliant attributes and powers against the impotence of statesmen hedged in by vested interests, and timorous about new, imaginative departures in diplomacy. "The highest desires of the noblest hearts and best brains," he wrote in 1933, "do not find expression through the methods of statecraft." Young people were ready indeed to make sacrifices. But, said Les, "there is no government in Europe with a programme fine enough, noble enough, or bold enough to express the idealism, enthusiasm and determination of modern youth. Youth today is just longing for leadership . . ." He added a statement that he wished to make to diplomats, which sounds remarkably like that of the leftist intellectuals of the decade: "If you are going to quarrel among yourselves you must do your dirty work yourselves, for we have more in common with the youth of other nations than with you. We have tried the method of war because you told us it would work and it doesn't work. We refuse to try it again. We have seen the graves in Flanders and Gallipoli, in Palestine and Mesopotamia, and the leaders we need today are buried there." He went on to make the suggestion that church statesmen in England, Germany, France, and Italy should initiate an International Christian Party—"The youth of the world awaits the clarion call of world-leaders who are international Christians."

The idea that politicians, unimaginative, wary, and blind, would resign their powers into the hands of the young, bright-eyed with principles and ideals and quite innocent of the hard unresisting facts in this pragmatic pig of a world, was something of a dream. And yet, thirty-five years and some four or five wars later, it has all fallen out somewhat so: no power has been gracefully resigned, but young men and women have made their voices heard and they have made sacrifices and, if they did not prevent the war in Viet Nam, they have caused it to come grinding to a close. Long before its so-called end, however, Les, then in his seventies, still abetted the young, his faith undimmed: "If I had a teenage daughter," he wrote, "I would rather she went to Viet Nam—yes, mini-skirt and guitar and boy friend, the lot—to minister to little babies, blinded, burnt and mutilated by damnable American bombs." Aged himself, by no means unmindful of all the folly and errors of the young, he looked to them to redeem things.

In the meantime, however, there were the other wars. In the summer of 1935 anxieties about peace began to focus on Abyssinia and the Italian threat hanging over it. The story of how high-minded sentiments and noble gestures mingled in England, in this lugubrious chapter of history, with half-hearted and impotent measures is well enough known: in August the Cabinet declared it would honour its obligations to Abyssinia, a fellow member of the League of Nations, and in September Sir Samuel Hoare, the Foreign Secretary, made a speech to this effect in Geneva, which was followed immediately by a show of naval strength in the Mediterranean. Anthony Eden proposed sanctions—the cutting off of financial aid and economic supplies to the aggressor. But policies motivated by honour were tempered by considerations of self-interest, by fear, and by the strong mood of pacifism surging through the country. The fleet lay idle at Alexandria while Italian troop transports moved through the Suez Canal bound for the African continent; and although, in the end, sanctions were applied, the list of interdicted commodities excluded oil, the one serious military need that Mussolini could not fill from within his own frontiers.

In August, in an atmosphere charged with apprehension, Les offered to a newspaper a suggestion that seemed to a few, magnificent; to some, curious; and to many, mad:

> Is not Italy [he wrote] pursuing the very policy which gave Britain some of the most desirable parts of her Empire? Are we not hypocrites in branding Italy, unless to all our prayers, we add deeds? To my mind we are in the position of a successful but unbranded burglar, now converted, but who has never made restitution, who expresses the greatest horror at the thought of a relative's openly declared intention of stealing a watch . . . Let us cease talking about sanctions, armed or economic . . . Britain should ask Italy what territory now British, would help her in her legitimate and obvious desire for extension, and then make her a definite offer of such territory.

Correspondents were not slow to raise objections. The suggestion was the spontaneous expression of a man of goodwill, of, indeed, a Christian, who measured immediate issues against the timeless spectrum of evil and good; it wasn't very politic or strategic, and it contained not a little misinformation about British history: Britain, Les was instructed in a letter to the editor, had in fact in the past returned a great deal of territory that it had held: to France, Martinique, St. Lucia, Guadaloupe, Goree, and her places in India; to Spain, Cuba and the Philippines; to the Dutch,

Java and the East Indies; to the Greeks, the Ionian islands; to Egypt, Egypt, etc. Other correspondents asked where was the territory that would satisfy Italy's needs and where was the population that would prefer control by Italy to control by Britain? And if it was to change hands, why should it not revert to its original owners? There was much opposition: as usual serious practical difficulties rose up between the vision and its implementation. Les had not really thought out all the political implications of his recommendation. But he was heartened in the debate by support in a broadcast from York in which the Archbishop declared that a redistribution of territory would have to precede a general peace. George Lansbury had written to *The Times*, "If the Churches, led by their leaders, will take the field, and tell statesmen the solemn truth that Christ's teaching and the teachings of all great philosophers is not moonshine, not sentiment, but cold solid truth, a great revolution in men's thought and action will take place." And Les was responding as a Christian leader with a Christian response, which to the Greeks and the rationalists looked like foolishness.

When on August 24th the Abyssinian war broke out, he once again had suggestions which he repeatedly offered in pulpits and newspaper articles: the Church, he proposed, should undertake to care for the wounded on both sides; it should organise a mass vote of all its members, "condemning aggression and offering to adjust the conditions that have led Mussolini to take this terrible step", and carry the result of the vote to both Mussolini and the Pope; Britain should call on Abyssinia to refuse to fight and to evacuate the whole population to British Somaliland. It was this last proposition that once again amazed his audiences and readers: eight million people were not, they protested, to be simply moved across the map; it could not be done. In fact, however, in the late World War more millions than that were transported to their various deaths about the earth. But Les did not consider the hard pragmatic problems of transportation, supplies, quartering and alimentation: he simply lit upon and proclaimed what he considered the only possible way of saving hundreds of human lives.

While the elements were gathering for the great storm, my father was developing his interest and practice in psychology. It had been sparked off, as we have seen, in a base hospital in Mesopotamia. But then, when he returned to Madras, Les had joined a small group of doctors and ministers who took it in turns each week to introduce and discuss a new book on healing. He was quite convinced that psychological healing was an important part of his ministry—or anybody's, for that matter: there was, primarily, the precedent of Christ, who had healed sick people who, for the most part, Les believed, were suffering from psychogenic illnesses, physical afflictions arising not from physical but from mental causes. The healing miracles, Les held, were not the exclusive prerogative of Christ, who had access to powers which we ourselves might command. He thought that the Christian church had lost the secret of spiritual healing at the end of the third century but might still recover it. Sometimes ministers and healers lit by chance upon their own healing powers and produced so-called miraculous cures. If Christ supplied the great precedent for Les's healing practices, other justification came from entirely different sources: Sigmund Freud held that psychoanalytic therapy was the proper sphere of both the educator and the pastor, and Carl Gustav Jung declared that it was the pastor who should be concerned with spiritual suffering. Then there was John Wesley, who commented in the *Journal* upon a woman who suffered physical pain which, arising from her distress at the death of her son, was unrelieved by the prescribed drugs: "Why then do not all physicians," he asked, "consider how far bodily disorders are caused or influenced by the mind; and in these cases, which are utterly out of their sphere, call in a minister?" Les anticipated with some longing the day when medical doctors and ministers would work together in healing; and there is no doubt that by his exertions he advanced it. He had the most sanguine expectations for the future use of spiritual healing: he believed, for instance, that mental methods would finally invade the province of material medicine so deeply that drugs would fall into relative disuse and some kinds of routine surgery would come to be looked at as mere crude carpentry. The future was to be quite staggering.

Meanwhile, the psychological healing that Les practised and discussed in lectures, sermons, articles, and books was firmly grounded in Christianity; it was predicated, one may say, on his frequently repeated belief that God's will for man was perfect health of body, mind, and spirit, and that to accept the values of Christianity and abide by its principles was a practical means to mental health. In *Modern Man in Search of a Soul*, Jung had written:

> During the past thirty years, people from all the civilised countries of the earth have consulted me . . . Among all my patients in the second half of life—that is to say, over thirty-five—there has not been one whose problem in the last resort was not that of finding a religious outlook on life. It is safe to say that everyone of them fell ill because

he had lost that which the living religions of every age have given to their followers, and none of them has been really healed who did not regain his religious outlook.

A few years after coming to Leeds, Les began giving lectures on psychology. He travelled about the county, going first to places with names like Heckmondwyke and Clitheroe and Osset, lecturing in little halls where the rain beat noisily on the corrugated iron roofs. His talk filled a gap in a series, perhaps, between a lecture on the Nile and one on how to spray fruit trees, or the life-style of the killer whale, illustrated with slides on the magic lantern, he himself, with his characteristic genius for lucid exposition, presenting the recently charted and still fabulous territories of the mind to the intelligentsia of the county borough, who had brought their husbands and their knitting. Now, in memory, I can see him still, some dark evening, belting himself up in his motoring clothes—a great mackintosh and gauntlets with red backs and white palms for signalling (no mechanical lights then) and going off into the rain, out to some little chapel in the wilds somewhere, to tell them about the inferiority complex or the power of autosuggestion.

Lecturing once on the way the mind caused apparently gratuitous lapses and idiosyncrasies of behaviour through repression, Les was asked by the secretary of the group who had invited him, "Why, then, is it that I repeatedly forget your name?"

"Whom do you hate most in the world?" Les asked.

"My boss," replied the man without hesitation.

"Why?"

"Because instead of coming and facing you with his complaints like a man, he sends you a note in that sneaky green ink of his."

"You forget my name," explained Les, "because I use green ink and you associate me with him."

He was not for long to deliver his message in the relative obscurity of the market towns of northern shires. He became well known as a psychologist, in part through the increasing power of his church sermons and in part through his regular articles on psychology in the *Methodist Recorder* and, after 1929, through the gathering of these into a volume called *Psychology in Service of the Soul* which had a foreword by Maxwell Telling, Professor of Medicine at the University of Leeds, who had been of help to Les in his psychological work. Soon he was speaking in cities and in large halls—Queen's Hall in Blackburn, the Y.M.C.A. in Halifax; the City Temple in London. And they were filled.

He had, of course, as we have seen, an extraordinary gift for speaking, and his ability to bring these mysteries of the soul into easy acceptance by his audience was part of that enviable talent. At the same time the subject of psychology lent itself to popularity: here the perverse behaviour of men, women, and children was rationalised; incurable illnesses were understood and became curable; obsessional fear, involuntary hatreds, lapses of memory, were fitted into patterns obedient to law. And much of his illustrative material Les derived from his own experiences in healing.

He had a set of rooms in the building opposite the chapel in one of the small streets of the city, presided over by the chapel-keeper, a man called Pickles, an ex-sergeant-major of an infantry regiment, with a waxed moustache like Kitchener's that came to a point on each side like a gimlet. Every afternoon of the week Les would drive down to these rooms and interview people with widely varied troubles. They were all people who had been referred to him by doctors, who in many cases had given them up. He took on patients in those days whom later he would himself have referred to psychiatrists; but back then in the twenties, psychiatrists were rare birds, and if he himself had not tried to help those who came to him, nobody would. Coming to his barely furnished rooms in Leeds were men and women, physically healthy—tough even—who were on the brink of despair, or men and women wasted by no known disease, whose afflictions neither drugs nor surgery could alleviate.

He made a practice of trying to learn about people before they had begun to speak. He said he had derived the habit because he had read that Conan Doyle would practise making observations in trains, trying to see what he could learn of people and their personalities by watching eyes, mouths, fingers, and bodily movements. Les watched people as they came into his rooms and made deductions from the manner in which they held umbrella or cigarettes; he noticed the strength of a handshake, how people sat down and how completely or incompletely they gave themselves to chair or couch; he noticed particularly give-away behaviour—how a person might twist a wedding-ring or how his rate of breathing might increase as he talked about certain subjects. His regular procedure was to have the patient lie on a couch with all his muscles relaxed—the situation dear to *New Yorker* cartoonists—and to let him talk. He would ask the patient to tell his life story, beginning with the earliest memories and coming down through childhood, schooldays, adolescence, love affairs, marriage, career details, to experiences of the present. He would ask in particular that the patient speak of his fears as a child, his memories of these terrors, and of his special delights; what things he loathed and what things he was ashamed of; religious experiences and ideas, sexual memories, frustrations, and things envied; dreams. The aim was to release the memories bound in the unconscious mind, that dark agent which may control such physical functions as the breathing, the circulation of the blood, and the digestion, which makes our choices for us without knowing it, determines our habits of thought and action, has power over our behaviour, and often enough indeed is the terrible lord of life and death.

In recounting his memories, a patient would be encouraged to associate freely, letting one thing lead gratuitously to another. Or Les might himself offer a series of stimulus words, instructing the patient to say the first thing that came into his mind. By this method his hang-ups might be detected. Sometimes the patient would attempt to obscure the procedure by giving not the first word that came to mind but one less revealing; then the time-lag between the stimulus and the response would itself indicate that a sensitive area had been touched on. Often an intelligent patient, by talking about himself along directed lines, would himself discover the source of his own ailment. Les once treated a young woman of twenty-three who had come to find great difficulty in eating, and in two or three years had declined from a state of robust health and beauty to become hollow-cheeked, pale, and thin. She was a typist, and she had recently been jilted by a curate. She knew she ought to eat, but when she was confronted by food she felt what she described as a strong inward power prohibiting her. As he listened to her story, Les recognised the familiar and dangerous complex in which the patient, in the unconscious mind, wishes to do away with his own life. He explained to the girl in easy terms how her unconscious mind had sought to respond to her broken dreams and her hurt pride: it had said, in effect, "Life is over for you. Be a martyr. Offer yourself up on the altar of unrequited love. Fade away and die. Then you will punish the curate, and all the neighbourhood will pity you. 'Poor dear!' they will say, 'she died of a broken heart.'" The therapy took longer than this summary suggests: but the girl was made at length to understand how her mind had commanded the workings of her body, and how the inhibition about eating had arisen. And with this understanding being made conscious and with no further treatment, the inhibition was completely removed: she began to eat, gained weight and resumed her familiar features. Another case that came to his rooms involved a woman who became hostile and indeed violent towards her husband whenever he had to go to bed with an ailment. By a process of free association Les brought her mind back to childhood when she recalled the trauma she had incurred on learning that her mother had given birth to a son. Her mother in bed was thus associated with a sharing and consequent diminution of love. Her husband, as she recognised, had been a mother substitute—she thought of him, that is, not as a mate but rather as a protector; and thus his going to bed sick she unconsciously related to loss of love, and she became hostile. When she understood the mental processes involved she became completely cured.

Very often, discussion with a patient was not sufficient to lay bare the workings of his unconscious mind, and then Les would sometimes resort to hypnotism. To induce this condition, he used to bid the patient relax on a couch and try to relax mind as well as body. He used to suggest that the patient imagine himself lying in heather on a hillside on a warm day, smelling the wild thyme, the heather blossoms and the bracken, hearing the hum of bees, and watching small white clouds slowly drifting across blue heavens. He would tell him that he felt a wonderful sense of peace and contentment flooding his mind, that he was becoming sleepy, that he could hardly keep his eyes open, that his body and limbs were saturated in a most satisfying weariness, and that in a moment he would be asleep.

Hypnosis is a psychical condition in which conscious control by the mind is inhibited so that suggestions are accepted more easily by the unconscious, and ideas and impressions that the mind has long retained in its depths can be more easily dismissed. In this state the patient is in the control of the hypnotist: he cannot move his muscles without the latter's command; he no longer relates to his normal environment, but only to the suggestions of the hypnotist. He can be made to accept illusions, and he can be given instructions to carry out at once or, if so ordered, after he has awakened. One patient whom Les treated, a young girl of eighteen, was very suggestible and extremely easy to hypnotise. Her parents believed that Les had saved her life, and they and the girl herself encouraged him to make use of her in experiments and demonstrations. She became his favourite model when he showed the phenomena of hypnotism to clergymen and doctors in Leeds. Under hypnosis he told her to write her name: she rose, sat at the desk, and wrote her name. Les said then, "Haven't you heard? The letter 'e' has been excluded from the language. Try and write your name again." The girl complied, omitting the "e's" in her name. Les was able also to raise and lower her temperature to 104 and 96 degrees respectively, simply by suggesting to her that it was rising or falling; he was able to control the speed of her heart by suggestion; and, with a word, he could anaesthetise any part of her body. Once he touched her arm with a pencil, telling her it was a hot iron: she winced and shrank away. The arm was then bandaged; and then, in twenty-four hours, when the bandage was removed the skin showed a sear exactly as if it had been burned; she had of course been told that she would feel no pain. The same girl could be made to hallucinate: given a blank card and told it was a photograph of a person, she could describe features and comment on clothes. She could also be made to dream while in the hypnotic state: she was told she was dreaming that she was in a small boat and that the boat capsised. She had been gripping the sides of the couch; then she rolled off it, crying out with alarm. When she came out of hypnosis, she recalled the dream just as it had been given to her.

The almost absolute power the hypnotist apparently exerts over the patient—over his will and thence over the very cells of the tissues of his physical body—can hardly be contemplated without wonder or without some sense of the immense possibilities in medical practice and therapy that the technique might possess. Hypnotism had been used with great success in managing pain in childbirth; Les himself had used it in place of an anaesthetic in the extraction of a patient's teeth; but what engaged him most was the hope that, when the curious powers of the hypnotic state became more fully known, they might be enlisted to cure such maladies as cancer.

The memory is supposed to retain all experiences since birth, and under hypnosis it is possible to recover memories so old that, in consciousness, they are virtually forgotten. On occasion my father used hypnosis simply to aid memory. Once a widow came to him in great distress because the letters of her late husband had been inadvertently burned and she had forgotten their contents. Under hypnosis she related many details about her husband which she had "forgotten", including what he had written in the letters. Les took these down as she talked and then gave her the record when she awoke. The restocking of her mind with these things constituted a lovely enrichment of her life. Most often, however, hypnosis was therapeutic: when the contents of the deep mind were revealed and the cause of disability became known, symptoms vanished. One woman who sought help from Les used to lie awake nightly until three or four each morning and then, within an hour of falling asleep, she would dream vividly of a man coming at her as she lay in bed and grabbing her by the throat. She would fight him in the dream and then would run screaming along the bedroom corridor until she awoke in terror and panic. Les hypnotised her and she recovered a memory that had been buried for sixteen years: at the age of twenty, in the home of friends where she was staying, she had been assaulted by a man who approached her in her bedroom with outstretched hands, threatening to strangle her if she should move or cry out. It will seem incredible that such an experience could be forgotten; but the natural tendency to scream and run (together no doubt with her own unconscious desire for sex) had been completely repressed below the conscious level of the mind, on account of the man's threats, to be expressed for years afterwards in the dreams; for in dreams, universally, desired behaviour is often enacted. Under hypnosis as she expressed the forgotten horror, she screamed repeatedly and thrashed about violently under Les's restraining hands. As is frequently the case when the recovery is attended by such violence, the woman was completely and immediately cured. Awaking from hypnosis she sobbed a little, and then gathered herself together and left, unafraid and in her right mind.

Sometimes Les was able to penetrate to the perilous stuff in the deep mind only by interpreting patients' dreams. The manifest content of a dream—that is the images that we experience when dreaming—is an expression in symbolic form of unconscious desires and fears, sufficiently disguised to ensure their concealment from the dreamer. One patient who came to Les's rooms in Leeds brought a message from her doctor saying she was on the verge of a severe nervous breakdown which he was powerless to forestall. She was indeed in a piteous state of anxiety, and yet in the free associations of her talk she revealed little. Not until he had interpreted a dream that she reported did Les get a glimpse into her troubled mind. She dreamed that she stood in the porch of her house while in the street a great storm raged. Her brother, bareheaded and without raincoat, stood in the street in great distress. At last she rushed out, covered him with a coat, and brought him in, getting drenched herself in the process. When she ended her account of the dream, Les told her gently that she must at once write to her brother and make up their quarrel. She was nonplussed at what seemed to her an uncanny knowledge of her affairs. But the dream was in fact fairly straightforward in its underlying meaning. The brother, who appeared as himself and not in the form of a symbol, had done something that had brought a storm about his head and brought his sister to quarrel with him. She was preserved from the storm, in the self-righteous conviction that she had done nothing wrong and need not leave the shelter of her own house. At the same time, by her conscience, she knew that she *ought* to succour her brother, and hence in her dream she symbolically satisfies this moral imperative by bringing him protection. After the symbolism was explained to her she followed my father's advice, and within a week the symptoms that had defied medical remedy had disappeared.

That was the sort of thing Les could do. He had hundreds of cases. Through England and, later on, beyond it his reputation as a healer spread, and people came from everywhere to see him in the rooms in the back street in Leeds. Day after day, after a morning in the study, Les would go spinning down to the rooms in his little Morris car to interview people in varieties of distress. Some would return a few times, others repeatedly for many hours until the great weight that crushed the mind had been prised up. He was scrupulous about the use of the word "cure", being constantly on guard lest what might appear as a cure should prove in fact to be a mere relief of symptoms: he was equally scrupulous in disclaiming his own function in the healing process, pointing rather to the therapeutic powers of God focused by the Christian religion. He never took a fee. Once a grateful man whom he had healed gave him a great oak desk with a roll-top which he used for the rest of his life. And once a rich lady whom he cured, who must have spent hundreds of pounds on doctors, gave him a small brown-paper bag with two peaches in it! In those straitened days, peaches were unknown to our family, and Les carried them home as a prize. They turned out to be rotten, however.

5

His work among psychological patients in Leeds confirmed Les in the idea that the mental sufferings of men and women arise most frequently from buried experience: they have done things or things have been done to them the memories of which they have thrust down into the unconscious and, to all intents and purposes, forgotten. These repressed memories, these secrets that people preserve from themselves, may then become the sources of the anxieties and neuroses which the psychiatrist is called upon to treat. Much experience is repressed because men feel guilt about it. And of all our experiences that generate guilt, those related to sex are the most common. Les found the sources of many anxieties where Freud's teaching had led him to seek them, in maladjustment to sex, in the repression of sexual desires and fears in childhood, in the unconscious concealment of early sexual experiences. Sometimes, only too often, a patient had been the childhood or youthful victim of a pervert and had been driven by threats into silence and secrecy. Eighty per cent of the people who came to him for help were in difficulties that derived originally from sex.

In every civilised society, and indeed in primitive tribes too, sexual activity is hedged about with taboos or magical rituals: it is something special and mysterious. Les felt, however, that a great deal of mental distress could be alleviated or could at least have been prevented if men and women could be persuaded that sexual activity was not in itself shameful; and he sought eagerly in lectures, sermons, and writings to disinfect the whole subject of the guilt, fear, and horror that have traditionally attended it. He was not, that is to say, permissive; he believed very strongly in self-control, continence before marriage, chastity, and monogamy—the whole works; and later, after the Second World War, he could be sharply dismayed at the casual fornications and the licentious antics of the brave new world, the pornographic novels and movies, revues where the players take off their clothes, and poets who expose themselves. In the twenties and thirties, however, his aim was to relieve people of the pernicious belief that sexual desire or its consummation were in themselves morally wrong and culpable, and to remove from them the sense of sin that arose from instincts inherent in human nature, God-given, which could not be denied simply by pretending they did not exist. In lectures and sermons he repeated this message: he explained how sexual desire was not to be repressed but recognised; and he described how it might be sublimated, its energies geared to all kinds of endeavour, acceptable to society and satisfying to the individual spirit.

He was also anxious to disseminate knowledge about sex. People who were young adults and adolescents in the nineteen-twenties had received very little education in the subject, and Les knew how much mental suffering was attributable to their ignorance. In these days, when children receive sex education in school and are able, by the time they are twelve or so, to discuss perversions as glibly as their parents used to enumerate the causes of the Thirty Years' War, it is hard to imagine the profundity of the ignorance about sex that obtained even among intelligent men and women some fifty years ago. One young woman Les knew had slept around in hotels with a man she was fond of, not considering that it might be wrong and having no clear understanding of possible biological consequences, but only thinking it a rather grotesque procedure and a curious disposition in her friend that he should wish so frequently to engage in it. Once, however, when he kissed her with unusual violence, she was afraid she had been made pregnant. This woman was an honours graduate of a university, and her parents were Christians. Such naïveté, even in the twenties, is scarcely credible, but it was not unique. Another young woman Les knew had been out to dinner with her uncle, and he had laid his hand on her shoulder. When she returned home she declared to her alarmed mother, "You simply must tell me how babies start, because I think I'm going to have one!" This woman too came from a Christian home. Another woman, sent to Les for psychological help, was chained in a marriage she hated because years previously the man had touched her breast and, expecting to get pregnant, she had married him against the eventuality. Another woman he interviewed had made an unsuccessful attempt at suicide, because she thought her first period was a sign of venereal disease.

He found that ignorance about sex, amazingly prevalent everywhere, was especially so among religious people who had been brought up in hushed piety, the simple details of human physiology left unwhispered. Les himself had had little enough instruction from his father in these things. Once there was a report in the local paper of a boy the family knew who had been ordered by a magistrate to pay alimony: Leslie's father, that shy, good man, tossed the folded newspaper on to the dining-room table where his son was doing his homework and remarked, "I suppose you know what that's all about." "Yes," said Les, equally shy, having no idea at all. His mind was filled with the mystery: what monstrous thing could the boy have done to the girl that his parents had to pay for? Something worse, he guessed, than pulling her hair at school or pushing her over in the mud, or even breaking the windows of her house with a catapult. He darkly wondered about it for days, too scared of his father to ask him outright, and having no one else to turn to. And the moment of fear and mystery in the dining-room, with his homework on the green cloth of the table under the gaslight, and his father rising from the deep chair, giving him the paper and leaving the room—this moment he never forgot.

With the aim of instructing people in these things and guiding them in sexual conduct, Les brought the subject up often in his public communications. Then in 1931 he published a book called *The Mastery of Sex through Psychology and Religion*, with some assistance from a medical doctor, Marion Greaves. It opens with a chapter demonstrating what evil has been committed in the name of propriety by silence on this subject, and describing a whole sheaf of sad cases similar to and including those summarised above. It outlines the necessary facts, warns of the dangers of disease, discusses the unwitting cruelties which young people inflict upon each other by casual petting. It discusses the unhappy marriage—causes and cures; it discusses birth control, the single state, masturbation, perversions, social attitudes, the curiosities of the law and other associated matters.

It is a good book. It has become dated, of course; *mores* and laws have changed, sex education and the wholesale lifting of the atmosphere in man–woman relationships have rendered some of the solemn passages of instruction a bit unnecessary. But it is clearly written in a sensible, no-nonsense style. No one today would hesitate to give it to teenagers to read if they could be got away from the television long enough, though, as I suggested, the teenager of the seventies would find most of it old stuff. Not so in its own day. None of my father's books, I believe, met with such demand on the one hand and such animosity on the other. Many were the letters of gratitude he received from the kind of people for whom the book had been written. Teenagers read clandestine copies, hidden from their parents, and passed them on to each other; and the publishers received direct requests for copies in plain wrappers. But for all the admiration there was also a good deal of opposition and anger. The bare presentation of such material, not in obscure scientific jargon but in simple terms that the uninitiated could and did read, was a fearful innovation, and that this should have come from a Christian minister was shocking.

The book was banned in the Irish Free State along with such other variously constructive treatises as *Midnight on the Place Pigalle* and *Lady Chatterley's Lover*. Most of the hostility came in private letters addressed to the author, but there were some hostile reviews. In one Australian paper the book was labelled abominable, "as violating the privacy of the marriage chamber". There was more responsible opposition, however, that suggested that the reticence that had hung over the subject of sex was not without purpose, not necessarily misplaced, and that taboos were not bad just because they happened to have been encouraged in Victorian times. But the great majority of the printed notices were very favourable; the work was considered bold but wholesome, and of extraordinary value to practising physicians, ministers, teachers, and the lay public.

Les believed that men and women react more violently to situations in which they themselves have hang-ups. A fellow minister protested that he would be horrified for the book to fall into the hands of his daughter and besmirch her mind, the daughter herself having recently written to Les to say that if she had only been given the book to read earlier she would not now be in the trouble that her parents' policy of ignorance and respectability had brought upon her.

While my father was in Leeds we spent our holidays sometimes in the Isle of Wight with my mother's parents, sometimes on the north coast of Wales, and sometimes at Filey, on the east coast of Yorkshire. At these two resorts my parents would rent furnished rooms, and each day they would make the safari down to the beach with us children and the gear—the toy boats, the bathing togs, the shrimping nets, the beach ball, the spades and buckets, and the sandwiches and tea-things—and then in the late afternnon carry, coax, or haul it all back again. These places were pretty good in the summer for people with children: the beaches had rocks and sand: they were good for swimming, wading, flying kites and sailing the little boats; children could be left to get on with it while parents fixed the tea or just lay in the sun with the *Daily Telegraph* covering their faces. At Filey, there was the Brig, a tongue stretching out into the North Sea and washed over by its tides. It had dark green, round slippery rocks and pools, marvellously deep, with crabs and sea anemones and other such hideous minuscule monsters, dear to the kiddies. In the town there were cinemas for when it rained, gift shops, picture postcards, ice-cream carts—the works: a happy little playground. But there were always people at these places—often crowds. And as we have

A family group: author, Mrs. Weatherhead and Dick behind; Les and Margaret in front.

In Melbourne, 1951.

Leaving for a preaching and lecturing tour of the U.S.A.

Three famous Methodist preachers: Les with, left, the Rev. Dr. W. E. Sangster, and, right, Lord Soper.

abundantly seen already, my father's strong instinct was to withdraw from human company and, as far as he could, seek out the lonelier places of the earth's crust—or at least such lonely places as lay within range of his little motor car.

While he was in Leeds, thanks largely to the sale of his books, he was able to buy a cottage on the coast of Sussex. As I look back on the transaction I see it touched at all points with the marks of my father's dreams and memories. It was for one thing somewhat remote, at the edge of a small village, the back-garden gate opening on to the actual beach itself, the shingle that was heaped up along the high-water mark. At night as you lay in bed you could hear waves breaking, and the roar as they sucked back the pebbles. The cottage was primitive and must have reminded Les, for some things, of One Barrow Lodge Farm where he had spent holidays as a boy. After he had bought the place, he found that the lawn was large enough for a tennis court, and so the poles and the wire netting were set up, and there were evenings of tennis, like the good old days in Egmore. He called the place Rowanwood; there weren't any rowan trees around, but that was the name of the house of friends in Farnham, where twenty years earlier he had begun his ministry in the old days which, looked back upon, seemed sunlit and happy, as old days generally do.

We had our summer holidays in Rowanwood for some years until there was built right opposite the small cottage a roadhouse with bars and dance floors and swing music wailing out over the breaking waves. Inside, it was all chromium plate and plywood, the kind of roadhouse decor characteristic of the thirties. And when dusk fell over the sea and the beach, the neon-lighting came on and the cocktail crowd came droning in in their sports cars—young men with loud blue suits and brilliantined hair, bringing their peroxide blondes, who quacked and giggled late into the night. It was a monstrous invasion, we all thought, of the decent darkness. Rowanwood seemed soiled. Shortly Les sold it.

There were good holidays there, though, while they lasted. The church was generous, and Les got six Sundays off in the summer—summers were longer and warmer then than now: we were in and out of the water all day long, it seems, under an endless blue sky. There was swimming or canoeing or tennis or fishing, or just messing around.

In 1934 Les took his holiday earlier. In May and June, thanks to the generosity of Ernest Appleyard and his wife, he travelled with them as their guest through Europe and Egypt to the Holy Land, Palestine as it was then called. They had a big car, a Morris Isis, and they drove through France, Switzerland and Italy, shipped the car to Alexandria, and then drove over the Sinai desert north into the Levant. They visited nearly all the places mentioned in the gospels, and Les was thrilled with the experience. What it meant to him to see these places was, in a way, similar to what his psychological knowledge had meant: both gave him new and realistic insights into the life and activities of Jesus.

It appealed to him immensely to see the actual terrain where Christ had been and to survey the sights He had seen. He was dismayed by the elaborate and often meretricious decoration with which a number of the holy places have been embellished: in the Church of the Holy Nativity at Bethlehem, he records, in the narrow cave, now subterranean, which is said to be the actual place of the birth, there were fifty-three lamps burning. At the tomb in the Church of the Holy Sepulchre in Jerusalem there were many more. In all the churches marking holy sites there were ornaments, some mere fustian and trumpery like Christmas-tree decorations, others of valuable metals and gems: there were stars and crosses, sometimes stuffed models of the Mother or the Babe or the Christ figure; and there were always lamps. In the Church of the Holy Sepulchre, Copts, Greeks, and Catholics vied in their ceremonial processions; sometimes where these coincided they would compete in singing their chants, the volume of their hallelujahs growing as they tried to outdo the opposition. As Les and the Appleyards "did" the traditional places, Arab guides regaled them with what they presented as unquestioned facts: "Here, please, Jesus our Lord Christ was born"; "Here Jesus Christ was buried"; "Here, please, from this very rock Jesus Christ ascended into Heaven."

From the locations of dubious authenticity and from the gaudy shrines it was satisfying to turn to the natural scenes, the Mount of Olives, Mount Tabor, the hills and the Sea of Galilee: the country "over whose acres walked those blessed feet", the sights Christ had undoubtedly seen. The mountains of Transjordania still turned purple in the evening and the Mediterranean Sea still sparkled. Many days of their trip were packed with fascinating encounters; what thrilled Les most, however, were the things that seemed to be going on in just the same way as they had done nineteen hundred years ago: glimpses he had of fishermen, for example, throwing nets weighted to close at the rim, or women coming to the well, or once, in Nazareth, a carpenter's boy helping his father build a house—just an ordinary child, probably, with bright curly hair who threw back his head and laughed, bringing to life vividly in a flash the boyhood of Jesus.

Things did not always work so harmoniously, of course: modern innovations had inevitably spoiled the picture in places; and some of the natives, who ought to have concentrated on being picturesque, showed an undesirable weakness for modernity and its conveniences. At one place the women came to the well carrying petrol cans which are a good deal lighter than the proper earthenware pitchers; Les noted in another historic spot a monstrous iron railing that ruined the scenery, and sometimes in villages there were children swarming around clamouring for tips, when for colourful effect they ought to have been herding sheep or following a plough, whistling psalms.

All the same, there was much that was unchanged, and from the land and the unchanging customs of the people and from local legend and anecdote, Les gained a better understanding than he had had previously of many passages in the Bible. There is, for one instance, a rather bewildering place in Matthew's gospel in which Christ tells Peter to catch a fish to see if the temple tax is in its mouth: "Go thou to the sea, and cast a hook, and take up the fish that first cometh up; and when thou hast opened his mouth, thou shalt find a piece of money: that take and give unto them for me and thee." This has been thought to imply a miracle or to mean, "Catch a fish and pay the tax with the price it brings." But when Les and the Appleyards were at the Sea of Galilee, staying at a German hostel, they were given for dinner "St. Peter's fish". Their host explained that this was the *musht*, the fish Christ referred to in the text, and that it had a curiously large mouth. Furthermore, like certain birds, it was attracted to bright objects, such as the half-shekel piece, the tax due, which were frequently found in its mouth. Les decided that Christ was not suggesting a miracle at all, nor that the fish should seriously be caught and sold: He was saying, Go and get a fish and see if there's a half-shekel in its mouth.

The party had been close witnesses of a storm on the Sea of Galilee. This body of water is surrounded by hills and mountains, and often winds are funnelled down the sides of the slopes on to the lake and, without a moment's warning, they whip up a storm which, equally suddenly, may be allayed. They saw this. And Les was reminded of the occasion when Christ stilled the waters; and he wondered whether the recorded words, "Peace be still", thought to have been addressed to the waves, were not rather directed to the frantic disciples, while the water, by concidence, at that moment immediately ceased to rage, a phenomenon familiar enough on this lake.

The sight of a sheepfold and shepherd vitalised the twenty-third psalm for them and some of its features explained things. But one of the most vivid sights that Les brought back in his mind to reconstruct in his sermons was that of the desert between Jerusalem and Jericho. We read simply the text, sufficiently colourless, that Jesus was forty days in the wilderness; but the text provides no image of that wilderness. Les found it terrible: it is covered with stones and is waterless; there are hideous dry ravines and bare hills carved by the wind into cones, and there are twisted yellow rocks. There is little life, only a few jackals, ravens, and hyenas. H. E. Fosdick relates the fate of two students who attempted to cut across this desert: they were young men in good health and they had only gone a few miles from Jerusalem, but one was found unconscious, and of the other only traces of his dismembered body were ever found at all. Only recently the world has learned how that same dreadful terrain took the life of Bishop Pike. Here was a place which to see with one's own eyes was to fill the word "wilderness" with meaning.

Much of the attraction of Palestine was that it made the New Testament live as a modern story. On returning, Les produced a book, *It Happened in Palestine*, in which, as he says, he has tried to use his experiences in the country "in the painting of word pictures which aim at making the figure of the Master stand out more clearly still". In many instances he has made biblical episodes come to life by realising them, with some imaginative re-creations, as modern. Landing at Capernaum in Christ's day "you would have seen rows of sheds, heard the sound of hammers, smelled the dye-works nearby, and the fish being gutted, seen the smoke of pottery kilns and read, possibly, among other advertisements,

ZEBEDEE AND SONS
FISH CONTRACTORS"

Les modernised the Martha–Mary dialogue at Bethany—the occasion when Mary talked to Jesus while Martha was fixing the meal. The servants were gone to a national feast, it is supposed; Martha stumps off to the kitchen and starts banging pots and pans around, hoping Mary will get the hint; she breaks a dish, gets hot, and then, dishevelled, flustered, and getting angry, she comes back and instead of berating Mary addresses her guest. "You feel her hands are on her hips," Les writes, "and her face is flushed and angry, and the tears not far away. 'Don't you care that my sister leaves me everything to do alone?' In our colloquial way of putting it, 'A lot *you* care that she leaves me to do everything.' She is blaming Him as well as Mary. 'Tell her to come and help me. We shall never get dinner at this rate.'" The book was fairly well received, even if some reviews were a bit guarded. As an exegete Les was found "somewhat wayward and fanciful" by one critic; he regarded Jesus as "the first Gentleman of the world" or "Big Brother" (this before George Orwell had appropriated the name) and not everyone would approve of his rewriting of scriptural conversations.

The party shipped out of Haifa and returned through Italy. On his return, Les summarised his experience: "The main impression that came to me out of those seven weeks," he said, "was the tremendous solidity of the foundations of our faith . . . a faith that is founded in history, in the life of a Person who spoke, and lived, and acted in places which may still be visited."

It Happened in Palestine came out in 1936, the year my father left Leeds. The eleven years he spent there spanned his thirties and early forties: and they were perhaps the most active decade of his life. He himself felt that his prime had been later, during his late forties and early fifties; and I think during these years a more mature tone and a greater degree of shrewdness become perceptible in his work. But the Leeds years were probably the most fertile: and there was not only unbounded energy in his work but also intellectual courage, and sometimes striking originality in his thought. In the eleven years he published eleven books. Three of them, including *The Mastery of Sex*, were on psychology. One was a thesis which had been presented for the M.A. degree at Manchester University, on the contribution of Victorian poets to the development of the idea of immortality. This now appeared as a book, *The After-world of the Poets*; "poets" and "Victorian poets" were then more or less synonymous to my father, and old Tennyson looms large.

One book of this period was an attempt to resolve the problem with which all men of the cloth must repeatedly come to grips during their careers, intellectually in the cool of the study and in the pulpit, but also more immediately in the press of grief, in hospitals, in rooms with drawn blinds, or in cemeteries in the driving rain—the problem of pain and grief. *Why Do Men Suffer?* was the first main confrontation Les made with this enigma: others were to follow, and I think perhaps his large usefulness to men and women, and accordingly his fame among them, lay mostly in his approach to suffering. Throughout his preaching career, it was a major theme. Indeed, for what cleric, one may ask, might it be otherwise, living in this dangerous century, knowing at first hand the pernicious effects of poverty, frustrated desires, disease, sin, and the manifold grievous products of wars? In the preface Les presents his book as the vision he has been able to realise after twenty years' thinking on the subject which, he said, had haunted him ever since he had begun to think for himself at all. He writes as if his ideas were provisional: but on the central issues I think he remained firm. What he declares in the thirties of the relationship of God's will to suffering, he maintained throughout the years: God's will for us is health of body, mind, and soul; Christ's way with suffering was to alleviate it; and in the Bible the woman who suffered was one whom *Satan* had bound. In a universe where men are free to commit sin and folly, suffering is accordingly free to occur. It is not God's will that a man should be killed in war or die by disease, but war and disease do not therefore defeat God's will, not ultimately.

Les did not believe that personal suffering was necessarily the product of personal sin, and he cited Christ's answer to the disciples who enquired of a man born blind whether he or his parents had sinned: "Neither did this man sin nor his parents." Sin, he believed, involved its own penalty—the penalty for lust was to be satisfied with lust; of greed, with greed; men were punished as their eyes were opened to their own previous short-sightedness. "And this is a different understanding of the phrase 'It is the will of God' from that which is popularly meant. It does not mean . . . that God hands out a cancer because a man has been lustful. It means that man is put into a universe in which he slowly and surely discovers that life will only work out successfully in one way, and that is God's way." This position he adhered to consistently throughout his ministry, through all his experiences of the suffering of the crippled, the dying, the bereaved men and women whom he was called upon to comfort. He held to it with his friend, Dr. W. E. Sangster, that good and gentle man, who, in his late fifties, through nearly two years' slow dying, bore with amazing grace the progressive deterioration and annihilation of his muscular powers, himself believing that in his own case illness was the fruit of his own sins and, as he termed it, "God's likes", the will of God.

Of the books of the Leeds period the most remarkable and yet the most characteristic, I believe, was *The Transforming Friendship*. Les had begun it in India, and then he had published parts of it as articles in one or another religious newspaper. The title epitomises the general approach: Christianity not as creeds and dogma and intellectual commitments but as a friendship with Christ. The book is divided according to various features of the transaction: the gift of the friendship, the reality of the gift, the intimacy, and so on. The chapters describe Christianity in the simple terms of an existential relationship, short-circuiting the approaches of the traditional churches. The whole volume treats religion as Les was wont to treat it in his books and sermons throughout his ministry: *The Transforming Friendship* is not indeed unique in substance among the thirty-five to forty volumes he wrote. But I single it out for mention partly because it was an early production and partly because it was most successful: during his lifetime Les saw it translated into ten different languages and sell over a hundred thousand copies in English. Quite properly the reviews emphasised the clarity of the book and the way it dwelt upon the simplicity of Christianity. It was designated "a young man's book"; but it contained, said one critic, "flashes which make the heart dance".

Les always declared afterwards that the Leeds years were the happiest of his life, adducing among other causes his own health and vitality, which after the thirties were never again unimpaired, the excitement of his work and its manifest success, and the warm-heartedness of the people of Yorkshire. It was, clearly, a situation happier and more propitious than the one in Manchester. For one thing the family was better housed. The first house we moved to in Leeds was a bit dark, and the neighbourhood was less than genteel. I remember there were cobbled streets and, after the rain, a mud puddle at the corner: it was a glorious puddle—I think it was the first real body of water I had ever seen, and we upended our tricycles in it and said they were ships; but it did little for the general environment. Neither did the man down the road who used to throw bricks and cobblestones at passing cars; he did it out of envy, my mother explained. Our garden wasn't much of a garden—just a few bushes blackened by Leeds soot, and no one ever went into it except to pass through to the street. But there was a big tree and in the tree there were always cats, many cats; and in the dark their agate eyes gleamed, and they howled at night.

Then after a few years we moved to a brighter place on the northern edge of the city, a large stone house with big windows, with open fields close by and, on the sky line, the Yorkshire moors. It was in Leeds that Lorna Margaret was born to my parents. And then in 1932 a boy was born dead. Les laid him in a box with some arum lilies and then carried him into the study and read the baptism service.

> I know many would say it was useless, [he wrote] but I believed he existed somewhere and that was all I had of him. I offered a little prayer, giving him back to God, and I put a card on his breast. "David Leslie, Darling Son of Leslie D. and Evelyn Weatherhead. Safe in the arms of Jesus." And there I truly believe he is or perhaps my dear parents are taking care of him for God. I am sure he is alive somewhere. He lived and was loved in a sacred cradle for months though we could not see him. The next day he was buried under a tree in the churchyard of Moor Allerton Parish Church.

The vicar had put on a cassock and read prayers over the grave. "I have no resentments, least of all against God," Les wrote. "I feel as if I had had him for a long time. I know I shall see him in the Morning."

Much later, in séances to which my father used to go, the medium would frequently say to him, "There's a boy here who says he knows you"—a boy of eight, or ten, or twelve—whatever it might have been. And Les would count the years since 1932 and deduce that it was David Leslie in the other world, where he was regularly reported as being happy, as far as I recall.

It was in Leeds, that "Gert" joined the family. "Gert" was Gertie Hutchinson, who came early in the thirties and served my parents for over forty years. They had had difficulty getting help: in Manchester, one woman said her husband required a statement signed by a doctor that Mr. and Mrs. Weatherhead were respectable people before he would let her work in Cecil Street. The maids who lived in came and went. My mother was exacting. She was never less than kind, but she was a scrupulous housekeeper, and things had to be quite right. And when the girls talked with the tradesmen at the door, she thought it boded the most fateful kind of disaster, and she strongly forbade it. Once, years later, when Gert had temporarily departed to nurse a relative, the person who came in as substitute was a member of my father's congregation. All the same, after a week or two in service, she told my mother she was a "cantankerous old devil" and promptly left. After she had been gone for a bit, however, she was guided by God to return and help my father at week-ends. So on Saturday nights she would enter the house with her week-end bag, take possession of her old bedroom and the kitchen, and move into action. Sunday evenings, she would clean up the kitchen and silently steal away. She must have had a fine spirit. My mother steered clear of her, knowing well which side her bread was buttered—knowing that it was pretty close to not being buttered at all! She was relieved when Gert returned.

Gert had the sweetest disposition imaginable; and it was permanent. She seemed to lack the ability to complain or to be ruffled; nothing could upset her—muddy boots on clean floors, unexpected visitors on washing-day, rain when she was planning to go out. All the hazards and demands that the life of an irregular family imposed together with the repeated defeats of order and routine she met with the cheerful characteristic phrase, "It doesn't matter." In a marvellous and selfless way she loved the family. To all appearances she has never grown a day older; and she never treats us as if we have grown older, either. When I was pushing fifty, when talking to me Gert was still referring to Les as "Your Daddy".

She has bright red hair and a pale complexion and when she is amused or surprised she puts the knuckles of her hand to her mouth and raises her elbow and looks as if she were in a conspiracy of mischief.

My father worked hard in Leeds, preaching, interviewing, writing, lecturing, and visiting the sick. He was outgoing: he had a cold bath every morning and did a few push-ups; he played golf on Mondays. But he never gave much thought to himself, to his own physical and mental well-being. He didn't really begin to consider his own health until, preaching one evening in the City Temple, he experienced what he later said felt like a kick from a horse in the back of his skull. In the Leeds days he took his own health for granted; all the energies of his mind were directed to his work and to healing or succouring others. He was successful. We have seen some examples of the cures he effected, of the large congregations, and of the sales of books. But there was this other thing, the personal affection, that makes itself felt in the letters left over from those days and in people's personal memories. There were many, many letters offering conventional thanks for sermons and articles, and some unconventionally intense with a personal ring to them: "I simply must let you know how grateful I am to you for the part you played in my life," one letter reads. Of one of his articles another letter says, "I have waited years for this help, and had it been written sooner it would have saved me an illness of fifteen months." Another correspondent writes, "As this wonderful experience has come through your sermons, books, and last though not least your friendship, I feel I would like you to be the first to know of it, and thank you from the depths of my heart for your patience and kindness." A girl from Australia writes, "If it be that God has some part for me to play in the destiny of my native land, I shall be glad that I have told you that it was through you He spoke to me most clearly." These are samples, more or less random. There were hundreds like them.

It is in such memories and in the lives of men and women, who knows how many, pervaded with decency and kindness, that the remains of the Leeds ministry of my father consists. It would not matter to him very much—it would not, that is, be a personal defeat—that the church is run down now, that the windows are covered with grime, that a tattered notice announces one service per month, and that the general desolation makes it impossible to believe that forty years ago you had to line up hours before the service to get a seat.

One morning as he looked from the pulpit of Brunswick at the people gathered for worship he noticed two men whom he immediately recognised as visitors from London. It must have given him a bit of a start, a moment of elation, no doubt, to look down between two verses of a hymn and see these men and know for certain that they had come to offer him the pulpit of the greatest Free Church in England. When the offer duly came I don't think it took him long to decide that the move to the City Temple, the great Congregational church on Holborn Viaduct in the City of London, was his manifest destiny, although he was overwhelmed by the offer and could not believe himself adequate to the demands or worthy of the honour.

He was to be succeeded at Brunswick, Leeds, by William Sangster, a prodigiously earnest man, whose preaching combined in an extraordinary degree the warmth of his human friendship and the outrageous other-worldly challenges of Christianity. The word "challenge" has been debased by small-town politicians and speakers addressing graduating students, but in Sangster's sermons challenges existed for real: it was possible to be absolutely nonplussed at the terrible obligations being laid upon one in such a mild and benevolent manner by this kindly man. He was a man of learning, and he was humble; and Les loved him well. In his last sermon in Leeds, he said, "There is nobody on earth I would rather have following me." He enjoined his people to be as loyal to Sangster as they had been to him. "Just come on the first Sunday in September," he said. "Do that for my sake will you?" A man in the gallery shouted, "Yes, we'll come." The atmosphere by then was pretty charged.

It was a cool July evening when Les made this farewell sermon. The service began at six, but some men and women had been waiting outside the church since three o'clock in the afternoon. By the time the doors were opened there was a long line, four-deep, waiting to go in. The church was tightly packed. People sat in the aisles, on the steps of the pulpit, and in the pulpit itself. The proceedings were relayed through loud-speakers to the outside and a large number of people sat in the stone-flagged yard under the trees, some of them putting up umbrellas against the rain. When my father rose to speak, three thousand people became very silent, waiting with expectation for the familiar silver voice that had urged, enjoined, comforted, delighted and astonished them week after week for eleven years. When he paid tribute to the "amazing courage of ordinary people", women began to weep openly. "I have seen men take the most crushing blows and be brave," he said, "and I have seen women go through hell and keep their courage." And then, his own voice faltering, he added, "And I too leave a tiny grave behind."

While he was speaking, a colleague at a chapel down the street was expressing gratitude for Les's ministry. If it was a generous summary, it was not therefore inaccurate: "He has taught the youth of the English-speaking world," he said, "to think clearly and live purely. By his simple, straightforward and challenging speech he has interpreted the Christian faith so that the modern person may see the positive claims of Christianity."

After the service at Brunswick, after the last communion, and all the farewells, the handshaking, and final tender good wishes, when he left the church at last, an old woman grabbed his hand and said, "I'm a Roman Catholic, but I've prayed for you every night for years." One girl of twenty-one, dying of tuberculosis, whom he visited as he left Leeds, said, "I shan't see you again, but I shall pray for you until the end." And another dying woman wrote to him, "I hope to die with my spirit alert—waiting, ready, and my last earthly thought to turn in gratitude to you."

6

At the end of the summer of 1936, after the long holiday in the Sussex cottage, the family moved into its London home. It was a brand-new house, made of brick, and it stood in a nice street in the agreeable suburb of Finchley, N.3, a respectable neighbourhood that housed besides ourselves a doctor, one or two people who owned small successful businesses, and a few single ladies and widows who were, as they say, comfortably off. The land in Finchley is fairly high and the air is good; in the mid-thirties there were woods and open fields within a block or two of the house. It was well out of the city, and yet my father could get down to the church in his Wolseley in about twenty minutes. The house had been given to him outright by a friend from Leeds. Les and my mother bought furniture for it and chose the wallpaper. They were now in their early forties, and for the first time in their lives they owned the roof over their heads, the chairs they sat in, and the board. The study was a large room on the third floor with windows on all sides. One alcove lent itself particularly to being designed as a small chancel, with an altar across it and a cross and Thorvaldsen's "Christ". Les had coloured paper stuck on the window to make it look like stained glass. There was a kneeling-cushion and a large copper water-carrier such as Arab women used to carry on their heads; he had brought it back from the Middle East and it stood as if left momentarily while its owner stopped to pray. That was the idea—a dramatic touch. The main part of the study was in green—green rug, green wallpaper, and an arsenic green telephone. All round the room were shoulder-high bookshelves on top of which were ranged trophies from here and there—brass Indian gods, a *kukri* from the old North West Frontier, a sword with an inlaid silver scabbard from the Arabs, an Arab coffee pot, pictures of friends, and a large water-colour of the sea.

By early September of that year, the children had all moved away, the boys to Kingswood School, in Bath, and Margaret to Queenswood, in Hatfield, north of London. So my mother and father and Gertie had the house to themselves, until a stray cat, subsequently called Smudge by my mother and a variety of less affectionate names by my father, one day wandered in and asserted squatter's rights. It followed my mother around everywhere, and Les accused her of loving it more than him. "If I lie on the floor outside the bathroom and purr when you come out, will you love me too?" he used to enquire.

On October 4th Les opened his ministry at the City Temple, where he was to serve out the remainder of his active career until, after twenty-four years, he retired from public life. In accepting the City Temple pulpit, which was Congregational, he had been most anxious not to forfeit his membership in the Methodist Church and had accordingly petitioned the Conference to permit him to retain it. Their response was more than gratifying: "I have seen the Methodist Conference in strange moods," recalled one member, "but seldom have I seen it so hearty and so enthused as it was when the request came . . . and when, with splendid acclamation, that great Conference gave its unanimous assent." Les was very moved when the President turned to him as he stood on the floor of the house and said, "Wherever he goes, Leslie will always be a Methodist."

He was now, however, to be the seventeenth minister of that Free Church which, when he began his service in it, was approaching its three hundredth anniversary. It had been founded in 1640. In this year, amid some tense excitement, the Long Parliament had confronted Charles I, and its assertion of power signalled to a number of religious Non-conformists that it would be safe for them to emerge from their secret meeting-places and declare themselves. One minister, Thomas Goodwin, who was later to become one of Cromwell's chaplains, had taken refuge abroad; now he returned to the city and founded a small church, a gathering of the devoted which was to be the ancestor of the City Temple congregation.

The most celebrated of the ministers who had served the church was Joseph Parker, minister from 1869 to 1902. In an age of great preaching he was marked as one of England's greatest preachers. He had a leonine head, a massive brow, and a voice like an organ. He was a master of rhetoric and histrionics, vehement, dogmatic, and assertive. His assertions were more memorable, of course, when they were wrong than when they were correct; one such occasion, on the eve of the Boer War, Albert Clare, historian of the City Temple, remembered as a boy. "In his inimitable fashion," says Clare, Parker said, "'On my way here this morning I saw on the newspaper placards the words: "War Imminent in South Africa."' (Dramatic pause.) 'THERE WILL BE NO WAR!' I was very young and confess with some shame that I felt disappointed." His disappointment, of course, was short-lived. Clare records his satisfaction on another occasion when, in 1899, Parker delivered himself of some sentiments about the Sultan of Turkey. "People felt," says Clare, "that something had been said that cried for utterance. The world seemed a cleaner place for the protest." After the Armenian massacres, the ex-Kaiser had made a public declaration of friendship with the Sultan. "When I heard," said Parker in his sermon,

> that the Kaiser went to the East, and after a dinner—hear this, for there is no more solemn word in the speech of Christianity—when I heard that the Kaiser went to a dinner, and in an after-dinner speech said, "My friend, the Sultan," I was astonished, I could have sat down in humiliation and terror. The Great Assassin had insulted civilisation, and outraged every Christian sentiment, and defied concerted Europe. He may have been the Kaiser's friend: he was not yours, he was not mine, he was not God's. Down with such speaking! and let every man's voice be heard on this matter. So long as any man can say, "My friend the Sultan", I wish to have no commerce or friendship with that man. The Sultan drenched the lands with blood, cut up men, women and children, sparing none; ripped up the womb, bayoneted the babe, and did all manner of hellish iniquity. He may have been the Kaiser's friend, but in the name of God, in the name of the Father, and the Son, and the Holy Ghost—speaking of the Sultan, not as an individual, not merely as a man, but speaking of him as the Great Assassin—I say, God damn the Sultan!

Those were the days when a man could brand "hellish iniquity" as such without wincing, and a man with passion could find the rhetoric to convey it and utter it without shame. Parker had tremendous oratorical power, and when he lifted up his voice in the pulpit, men heard him. His memory was still green in the 1930s.

It was he who in 1873 had opened the building on Holborn Viaduct which came to be called the City Temple and was the church in which my father assumed his ministry; and when that building was gutted by fire in an air raid in 1941 there could be seen, through the burned-down doors among the slanting charred beams and debris, erect on its pedestal the bust of old Parker, the brow smudged and stained, but the cold marble gaze contemptuous of whatever evil evil men might inflict on the stones and mortar of God's house.

Parker was followed in the pulpit by R. J. Campbell, and he by Fort Newton, Maude Royden, and F. W. Norwood. Then came Les. At the induction ceremony held later that October, Norwood sounded what seemed at the time a curious note. He himself, he said, had never had a welcome meeting. He had had few letters. There were no laudatory press notices. The first year or two of his ministry were the loneliest of his life. He had had to fight his way through misunderstanding and suspicion.

> This Church [he said] is a church that will take all you have got and would take more if they could find it. This old City of London is a very impersonal kind of place, it will squeeze a man dry, and it will not be overmuch concerned about him when he is dry. So long as the flair lasts there will be no limit to the appeals that are made to you for the outpouring of the treasures of your mind and soul in speech, in literature, in all manner of advocacy, and then, when they have drained you dry, they may begin to say that you really have not got much to say after all.

It sounded a note in a minor key among the welcomes and congratulatory speeches. I wonder if Les remembered these warnings when in 1960 he retired from the church against some fairly strong protests and made much of the idea that it was better to quit when the crowd wanted you to stay than to stay when it wanted you to quit.

In the same ceremony R. J. Campbell said that Les had been called to London "in the providence of God", for never was there a time when "great preaching was more sorely needed". Well, it may indeed have been providential: the world was certainly due for some signal of God's cognisance of its distress. At the time Les came to London, the Emperor of Abyssinia was making his last appeal to the League of Nations; the Sino-Japanese talks had become deadlocked; Franco was marching on Madrid. And, suitably enough, Shaw's *Heartbreak House* was playing in the West End.

October 4th, Les's first Sunday, was one of those golden autumn days with sun and a slight haze and crisp air. A crowd gathered quite early, and not all could be admitted; but the service was relayed to the lecture hall beneath the church. A pew had been reserved for Lloyd George, who came in with his party in the first hymn; John D. Rockefeller Jr., who was not recognised (and who, it must be added, would not have wished to be), was accommodated with a kitchen chair in the lecture hall. My father preached on the humility of God. He opened with one of those vivid unconventional descriptions of a scriptural episode that he could do so eminently well: it was of the Last Supper, and with his knowledge of biblical scholarship, his familiarity with Eastern practices, his understanding of the ways of the human heart, and with some imagination he was able to offer a beautiful re-creation of the scene and the situation, as if it had happened just the day before, as if he had himself been there. He presented the scene—the upper room built on the flat roof of Mark's father's house, the curtains so hung as to secure both fresh air and privacy, the lamp suspended from the roof, the table only eighteen inches high, and the guests reclining, the meal secret, and the servants dismissed. Then he described Christ washing the disciples' feet and proceeded to contrast with this episode the human uses and manifestations of power and importance; he asserted the rights of the weak. Here and there, there was humour (next day the *Daily Herald* had a headline: "2,500 Loud Laughs in Church"): "If I were God, I wouldn't stand the indifference of the modern generation. I'd jolly well stir them up . . . I'd put a brass plate on the door and send a few angels around." He had been, he said, at a reception for the Prince of Wales: "Everybody who had a uniform," he said, "wore it—and, I'm tempted to say, everybody who had not a uniform wore it!" And then there were some rhetorical moments, like this: "We launched the *Queen Mary*. Oh, we felt awfully proud of that. And the other day God launched a new planet and nobody will know anything about it for twenty million years." Or there was this: "Badoglio [Mussolini's son-in-law, the Italian air ace and officer] knowing that all things had been delivered unto him through the might of Italian soldiers who fought a half-civilised people with tanks and bombing planes and mustard gas, proceeded to enter a capital in pomp and glory and demand the surrender of all arms and of every native. Jesus, knowing that all things had been delivered unto Him by His Father . . . began to wash His disciples' feet."

It was well that Sunday morning to be citing Italian fascism and to be talking of the vanity of human power, saying, as Les did, to his large crowd, "The world cannot be free of talk of trouble unless we all realise the right of the weak to live"; for not far away in the same golden sunshine, at the Tower of London, Sir Oswald Mosley's Blackshirts were on parade to greet *their* leader as he drove by in an open touring car in his black military jacket, the grey riding-breeches, the jackboots and the black cap, to receive the raised-arm salute.

Coming to London, Les seemed to be drawn closer to the national life and its affairs. He had, of course, spoken out all his life on national matters, especially as they related to peace and war. Now, with his large congregation in the city and his much larger following throughout the country, he begins to appear as a national figure. When he makes addresses in the provinces he gets larger headlines in the local press and sometimes even notices in the august columns of *The Times*. He is invited to luncheons to meet statesmen; he attends Buckingham Palace for garden parties along with my mother, tense and tight-lipped, wondering if what she was wearing was the right kind of thing. Repeatedly in the later thirties my father's name appears with those of other professionals and men of affairs as a co-signer of letters to the press on matters of national and international concern. In August of 1937, along with a dozen bishops and deans, he voices apprehension at the prospect of British recognition of Italy's conquest of Abyssinia; in the next year, after the Munich crisis, he joins in a letter appealing for amnesty for political prisoners in Czechoslovakia; he signs another letter appealing for funds for Czech refugees, and yet another proposing a new peace conference. Shortly before the Second World War he was one of a deputation that carried a petition for a peace conference to the Prime Minister, Neville Chamberlain. At the end of 1937, in a letter to the *News Chronicle*, he was proposed by one correspondent for "Man of the Year", while others suggested Lord Nuffield, the King of the Belgians, Lloyd George, Paul Robeson, Victor Gollancz, Pastor Niemoller, and Anthony Eden.

There were indeed stirring national events during Les's early years at the City Temple, and his comments upon them drew attention. One day early in December 1936 a deluge of gossip suddenly flowed from the press with all the energy of waters long pent up: there was a day when every newspaper carried stories of the relationship of the King and Wallis Simpson. My father's view of the royal family was, I think, that of the average Englishman—a healthy pride and respect; if it had been otherwise he would not so frequently have spoken of Christ as royal. His view of divorce was also, no doubt, average—a disapproval falling far short of horror. He had never had much hesitation in solemnising the marriages of previously divorced couples. George V he had admired as a devoted ruler; Edward, along with most of his fellow countrymen, he had liked.

The discussion between monarch and Prime Minister, and Prime Minister, cabinet, and parliament seemed to rush headlong to their grievous conclusion. On Friday, December 11th, the King signed the instrument of abdication, broadcast his farewell to the nation that night, and shortly, under cover of darkness, left the country. Les hesitated and debated in his own mind whether he should address himself to the subject of the abdication the following Sunday morning or whether he should not turn his people's minds to other things. It was perhaps presumptuous, he thought, to deliver his own sentiments on a matter pretty far removed from his area of professional competence. But he decided that if on this subject as on others he could "set [men's] problems in the perspective of the mind and purposes of God" he should try to do so. It was, after all, considered at the time to be the most grave constitutional crisis in English history. So at the last minute he changed his topic, and he spoke briefly about the role of the king in the English constitution and its great distinction from that of a dictator. He declared that either praise or blame for Edward would be inappropriate, but he made a great deal of the fact that the welfare of the Empire had been put above the happiness of one individual. Reverting to his favourite image, he said, "This stands out in my mind like a great mountain above all the swirling mists of controversy and discussion . . . Whatever may be our views on divorce, high idealism, as it is generally conceived, has been shown to count in our national life and to count supremely." He added some praise for Stanely Baldwin, who had acted, he thought, "as a wise Christian statesman", and other words which, as hindsight has revealed, were unduly kind. When the Duke of York was crowned George VI, he hailed the substitution, looking a little askance now at Edward, who associated with a coterie of the sybaritic rich, and with more satisfaction upon George, who had inaugurated a camp for boys.

As everyone knows, events far more grave than the abdication of the British king proceeded during the last years of the thirties to disturb the peace of Europe and to haunt the minds of English men and women. During the Abyssinian War and the re-arming of Germany in the mid-thirties, Les had maintained a pacifist position and had advocated disarmament. In the next five years, his attitude changed, as did that of most British intellectuals. It is worth observing the steps, because they were followed by many who acknowledged his leadership and by many more, I suspect, whose own attitudes were composed with the help of Les's public declarations.

As in Leeds, so now in London he continued to speak out for peace, and still protested that young men should identify with the young men of other countries. He was still haunted by the memory of meeting Martin Gerhardt, the German minister with whom he had so much in common—so much more in common, he confessed, than he had with most English politicians. Even after the occupation of the Rhineland in 1936, which signalled to some the beginning of Hitler's career of territorial aggrandisement, Les declared that he had "definitely finished with war" and had signed the manifesto of the Peace Pledge Union, founded by Dick Sheppard, vowing never to sanction war again. He did not repudiate the use of force if it could be used as the police use it, exerting it only on the criminal and then only to bring him to impartial justice. But war was not such a use, and he disavowed it.

He maintained this position through the next few years, the bleak history of which is now well known: Germany annexed Austria and then threatened the Sudetenland of Czechoslovakia. Late in September 1938, when the British were on the brink of war with Germany, Neville Chamberlain flew to confer with Hitler; and then, having sacrificed the Sudeten areas, returned to London to crowds who cheered him, out of their huge relief. From an upper window in Downing Street, he said he had brought back ". . . peace with honour . . . peace in our time". In the House of Commons Winston Churchill saw the solution otherwise: "a total unmitigated defeat"; but no one wanted to hear that version. The dictators exulted. But the dominant mood in England was relief: all that week we had thought we were going to be immediately embroiled, and we had been spared. Les had been due to broadcast on Sunday, October 2nd, but on the previous Wednesday he heard from Broadcasting House that "if war should break out by the end of the week" the hour in question would have to be given to the Archbishop. If the international negotiations still proceeded, the present programme would stand. The B.B.C. had not anticipated deliverance. No one had. And then Chamberlain came back from Munich with his umbrella, waving a declaration and saying it was all going to be okay. The nation relaxed. But for most men, the rejoicings were thin; they paused to reflect on the losses the Czechs had incurred to secure the peace and to anticipate what might lie not very far ahead. The Home Office did not stop issuing gas masks.

The Sunday following Chamberlain's return was declared a day of thanksgiving and, undaunted by heavy rain, thousands of people who had not worshipped publicly for years flocked to churches which had not for years had to accommodate such crowds. Hundreds stood and knelt in the aisles of Westminster Abbey; in the Catholic Cathedral and the Brompton Oratory the National Anthem was played, and the City Temple had a larger crowd than usual. Like most clergymen on that day of liberation, Les had a thought for the price being paid: "Let us not forget the Czechs," he said. "It is no day of rejoicing for them . . . God has given us our lives and property, and the lives of our loved ones, but never let us forget that the Czech people are paying part of the price of our peace . . ." He added,

> I do not know whether you share with me a rather uneasy feeling. Do you feel just a little as though you had made friends with a burglar, on condition that if he took nothing more from you, or your immediate friends, you would not do anything about his taking anything from

> anyone else? I hope the "burglar" has been converted. Let us believe in the conversion, but let us not forget the Czech people in our prayers.

The press reported this latter comment extensively: it was typical of Les to express the moral situation of international politics in just such a domestic simile. Twenty-five years later the passage was quoted on a B.B.C. programme which was designed to evoke again the feelings prevalent in England during the crisis of 1938. In his sermon Les also offered to anybody who should write to him that he would put them in touch with a German pen pal. This plan developed into his "Friendship Invasion", under which scheme people from the City Temple visited Germany in the following summer and made friends. His sermon concluded:

> This week we have been moved to the depths. There has been wrought for us a great deliverance. What are you going to do about it? Shall historians say, "If they had done something, war would have been banished for all time. It might have been the beginning of a new era, but they did not alter their practices. Hate, distrust, suspicion, fear and darkness fell upon the world once more"? O God, that we may be guided in the way of peace!

We know now, of course, that it was too late for that.

Early in November, however, Les delivered the broadcast that had originally been scheduled for Thanksgiving Sunday, and his message was again a pacifist one. "It may be remembered," he said, "that the Great War lasted 1,600 days. On each day 7,000 men were killed and 14,000 wounded. Nine million children were made orphans, five million women were made widows. Thousands were driven mad. Hundreds are still suffering in body or mind. And, *at the end*, a conference was held in which feelings were not less violent than they would have been if a conference had been held at the beginning." He again distinguished between the use of force by the police and the use of force in war. His position was much as it had been throughout the decade.

The very day following the broadcast, however, there began a series of events in Europe that helped to harden the minds of generous and optimistic men in England against the German regime. On November 7th Ernst vom Rath, third secretary of the German embassy in Paris, was assassinated by a seventeen-year-old Jew. The murder was followed shortly by a night of such terror and atrocity against the Jewish race as was hitherto unheard of even in Germany: some forty men and women were murdered, and twenty thousand were arrested; thousands of shops, synagogues, and homes were burned and looted, and many other various and bestial humiliations were inflicted. Iniquities vile enough had been perpetrated against the Jews on earlier occasions; these on the night of November 10th, however, were not the casual sporadic brutalities of individual thugs and sadists, but an operation approved by Hitler and organised and executed by his government. Like everybody else in England, Les was horrified by the reports, and on the following Sunday he interrupted the service to call for special prayers for Jews in Germany "robbed of their possessions, hunted from their homes, beaten, insulted, degraded, hiding in the woods in this bitter weather, crying for food and drink for their little ones, and longing only for death".

More outrages followed. On one day in mid-March in 1939 German troops entered Czechoslovakia, and at the end of that day Hitler was able to declare that that state no longer existed. It was not a very propitious time for my father, along with other members of a small deputation, to carry the petition to Chamberlain, signed by a million people, urging him to organise a new peace conference. In sermons now Les repeatedly addresses himself to the international anxieties. In March, he presents Hitler to his people as pathological. He describes the neurotic young man, explains his Oedipus complex, how he hated his father until he died, and adored his mother until she too died, laying upon him as she did so the injunction to be great: how he burned with a sense of personal injustice and a longing for power. Les asked his people to pray for Hitler.

He maintained his pacifist position through the darkening spring and summer of 1939. He organised his "Friendship Invasion" and men and women went in small groups to try to contact German citizens. There were to be some sight-seeing excursions, but plenty of uncommitted time during which they could encounter German people. Invitations came from individual Germans wishing to play host to the invaders, but Les decided not to accept private hospitality in case some indiscretion by the guests should endanger the host. He was himself to lead a party, but in late August the Foreign Office sent him a telegram instructing him to call it all off and bring his people home. Some of them had made a number of friends; they had exchanged gifts and flowers. They had not heard about the crisis.

In June, in a long essay in the *Methodist Recorder* entitled "To a Young Man Liable for Military Service", Les again lays out his pacifist argument. But now, after the additional thefts of territory on the part of the Fuehrer, the anti-semitic pogroms, and the dreary catalogue of broken promises and lies which made the prospect of fruitful negotiation increasingly unlikely, he emphasises the need of restraining evil and he plays down the differences he had earlier enumerated between police action and war.

Chamberlain had promised aid to the Poles if they should be attacked, but Hitler, though angered, was not deterred. On September 1st the *wehrmacht* swarmed into Poland, with what piteous far-reaching consequences the world well knows. Faced with the manifest and colossal evil of Naziism Les at last seems to have abandoned the position he had taken when he declared he would have nothing more to do with war. In mid-September he suggests that the need to restrain evil surmounts the earlier considerations of the wholesale injustice of war. "Is evil to go entirely unrestrained," he asks, "because we have perfected no way of restraint? If this dreadful method of war is the only method by which evil can be prevented from spreading, are we to eschew it and let the evil spread?" The rhetorical questions imply a reluctant commitment to arms. But still he adds, "If you can see the right and the wrong in that I should like to hear about it." And up to the last minute and even after it, Les went on calling for a conference of young men to arbitrate with words, before matters should come to be settled with weapons and their lives.

In November of this bad year, Les published his book, *Thinking Aloud in War Time*, which contained all the conflicting views and sentiments with which he had been wrestling for twenty years. There is no easy or clear conclusion: the issues, as they are here displayed, are between two shades of grey, not between black and white. But without any trace of chauvinism, without flags or trumpets, he shows that he believes in Britain's cause. He describes how, after he had spent the night thrashing out the problems of war and peace with Dick Sheppard, he had at last accepted the pacifist position. Then Dick Sheppard had said, "Leslie, I know I'm inconsistent. If anyone attacked my wife, I should sock him in the jaw!"

At about the time the family moved to London, Les purchased the Dial House in St. Margaret's Bay. My mother had been looking through advertisements for property in London, when she happened upon the description of this property in Kent. St. Margaret's is a village on the coast between Dover and Deal, right on the corner of England: for a half mile or so there is a break in the white cliffs, and there is a bay. Behind it the land rises up again to the level of the cliffs, and here at the head of this shallow valley stood the Dial House. It was of brick; surrounding it in its own garden was a small copse of pines and macracarpas.

From the balcony of the house you could look down to a grass tennis court and beyond that a pool with a fountain and lilies and goldfish. Les used to feed the fish with little pellets of bread, sometimes holding these under water in his fingers; most of the fish were shy, but one of the noticeably unshy and aggressive ones he named Hitler.

We had some fine family summers down there. My father used to write quite a lot, sitting at a table in an upstairs window overlooking the sea. He was fond of sea bathing and walking on the short rabbit-bitten turf over the cliffs to Dover, or in the other direction towards Walmer Castle, where from the cliff on a fine day you could see before you the whole coast of England turning northward, and, over the Channel, Cap Gris Nez and the buildings in Calais. One of his favourite occupations was to take the car to Fleet Farm in the flat country behind Sandwich and watch the marsh birds, the water hens and coots, redshanks and sandpipers, and the great ungainly herons flopping into their nests at the tops of tall trees.

St. Margaret's Bay lies opposite the Goodwin Sands and thus the shipping comes in very close to the coast through the narrow channel. My mother would sit and knit on the balcony of the house with a telescope by her, and when the ships emerged from behind the cliffs she would put down the knitting and study their names and flags and their ports of origin through the glass. Then at mealtimes she would retail the day's shipping activities. Tramps, tankers, ferries, the regular Dover–Calais boat and the Ostend packet, all would come and go under her immediate inspection. During the war a little fort was built on the edge of the cliff that my mother had habitually commanded with her telescope, and after the war Peter Ustinov bought it and made it into a house. Then sometimes my mother's telescope would move five points west from the shipping lanes, and she would report at lunch, "This morning four ladies went into Mr. Ustinov's house and two came out." At the east end of the Bay Noel Coward had a house which after the war he sold to Ian Fleming. But my mother couldn't see that from the balcony, and we never knew about the Cowards' or the Flemings' comings and goings.

Often in the late afternoon we used to play tennis. My father had a terrific serve, and when towards the end of the match the dew rose and the grass became slippery it was quite indomitable. Sometimes after tennis he would run and turn the fountain up as high as it would go and stand where the sea breeze blew the water on to the flagstones. I see him now, in his white tennis things in the cold, falling water, revelling in good health and vitality, then running up the winding brick pathway to the house to get towelled off before supper. They were good days.

As I look back it seems to have been a paradise. There were little birds in the garden—robins, and linnets and chaffinches—and thrushes and blackbirds used to pipe away in the trees, and overhead the gulls soared up like blades of light in the sun. At night their cries welled up from the cliff face; the lighthouse beam swept across the garden and caught the tops of the tall pines; from out at sea the South Foreland lightship boomed softly, and the lights of Calais harbour blinked across the water. It wasn't, I suppose, much more than an English house and an English garden such as thousands of people have owned and enjoyed. But for Les it was a place where he could satisfy his permanent instinct to get away, the yearning known to some extent by all men, for the lost impossible Eden, the innocent retreat to be withdrawn to from "the fury and the mire of human veins". Something like that, I guess, it meant to Les—as, for that matter, it did to the rest of us.

Not my mother, though. And when Les, with a bit of a flourish, presented her with the deeds of the house, she said, "Well it's nice, but of course it won't be any holiday for me. You'll all still have to be fed and looked after!" She could be like this: there was a bleak side to her, and in moments where sentiment was called for she could be arid and rational and refrigerate my father's enthusiasm with one cold blast.

My mother was an intelligent woman and witty, and she admired Les beyond words and loved him profoundly. Her love was almost completely self-abnegating: she thought of herself as his helpmeet and she consistently upstaged herself in his favour. When he left Leeds, in his last sermon there, he paid her a tribute: she had put the home first, he said, and had made it a peaceful refuge for him. She had certainly put him first in all her reckonings; she always did. Sometimes she would get mad and exclaim, "I wish I were an American woman. They're free!" (she knew only wealthy Americans, I think). But her bondage was only her own affection for her husband and family, coupled with her monstrous sense of duty. Her purposes were entirely wrapped up in him: he was her life, and her joys arose from his successes rather than her own. On Christmas night, each year, after the children had at last gone off to bed, my parents would sit down and do a jig-saw puzzle. Les would do the colourful things in the middle—the ship in full sail, the red tunic of a sentry, a coloured flag on a turret, whatever it might be; my mother always dutifully did the sky, each piece repetitiously blue, or the endless green grass, plodding away to provide the framework. At her best my mother would enclose Les in a protective shell of affection.

There wasn't always peace though. Like many ministers' wives, she had much to put up with from the adoration of her husband by young women. I suppose there is an unsubtle connection between the putative masculinity of God and the femininity of most of the church congregation—a supposition not currently in fashion. But if God's vicar is male and handsome, as was the case with Les, there is a good admixture of erotic love in the Christian charity of many churchwomen. Ministers' wives have this to bear, along with the furnished manse, with the purple peacocks in the curtains and the unmodish furniture. The wife of a minister I know was assailed one Sunday after the service by a powerful youngish matron who positively bloomed with affection for him. "How can you *live*," she demanded dramatically, "how can you *live* with Mr. Magee!" "Oh," said his wife drily, "it's possible. Just. It can be done." Probably many ministers' wives have wished or half wished that their husbands' worshippers could see the great man in his shirt, waiting for a button to be sewn on his trousers—an image corrective of undue idolatry. For his part, my father liked young women, naturally enough, and he invited them up to the house for Sunday supper after church.

My mother's lot was not easy. No idealist is entirely easy to live with, and Les always wanted things to be better than they could conceivably be. He was sometimes impractical, inasmuch as he didn't want to be bothered about the daily details of life. He was concerned with ethereal matters, the thoughts and ideas, the joys and demands of a spiritual world. In a number of respects he was a platonist, and his attitude towards physical necessities could really be quite cavalier. And so he sometimes lacked sympathy when my mother wanted to discuss with him the substantial concerns of day-to-day living—how the laundry had ruined a shirt, or how the greengrocer had delivered unripe tomatoes. He might be making a sermon in his study, when she would appear with a can of corned beef holding it for him to sniff and demanding a verdict as to its freshness, and he would ostentatiously mark the place he was reading with a finger and offer in silent protest a token sniff. It wasn't just a matter of being disturbed in his work: he simply didn't think these things were worth discussing. For his own part, he wanted to talk about altogether different sorts of things, or play Beethoven on the record player, or watch a sunset in peace. He was like that. He could be pretty unhelpful with knotty household problems. He assumed that somehow they should solve themselves in the ideal life: and he didn't want to take part in the existential side of living where the laundry goofed, tomatoes failed to ripen, and beef went bad. "Oh, I'll buy you a *bushel* of tomatoes," he might declare, anxious to elevate the tone of things as expeditiously as possible. You could see his point: he had enough worries. But you could see hers too: my mother didn't want a bushel; she just wanted to talk about the pound she'd got, and receive some mark of the kind of love that would identify itself with her concerns. And she had a will of iron. So there were times when they didn't seem to be reading each other, and my mother could puncture his romantic buoyancy with a remark like, "It'll be no holiday for me", when my father gestured largely and handed her the deeds of the house by the sea where we could all withdraw from London and get away from it all.

From here, generations later, the years of 1936 and 1937 seem to have been part of another world. As we have seen, my father was fully and fearfully aware of the fragility of the European peace; but he believed—pretty firmly, I think—that it would be kept; there would be no war. He was optimistic about his work at the new church. He had started no less than nine new organisations, such as the Samaritan League, a group of men who gave help to those on whom bad times had come; the Friday Fellowship, a group which met each Friday evening to discuss religious questions; the Psychological Clinic, a team of doctors who would give their professsional services. His own preaching was at its height. In the early spring of 1937 he writes that it had been one of the stiffest winters he had ever known; but he adds, "I have never known a happier winter." He had, as they say today, everything going for him.

At the end of his first year at the City Temple he accepted an invitation to lecture at McGill University in Montreal. It was to be a brief visit: he sailed on the *Queen Mary* on September 15th and, after the roughest trip that ship had experienced, arrived in New York on the twentieth and spoke to a packed audience in the Fifth Avenue Presbyterian Church. He caught the night train to Montreal, arriving on the twenty-first. He lectured that night and the three following days, gave a luncheon address in addition, and sailed on the twenty-fifth, arriving in Southampton on October 1st, full of beans and ready to preach in London on the third. It was a mad dash: it was the sort of schedule arranged by a man who values every minute of his waking life and doesn't have to think about his physical constitution. Les was such a man in those days at the beginning of his ministry in London. And in the summers at the Dial House, after tennis, when he would come bounding out of the cold fountain and up to the house, where the lights in the windows were golden in the dusk, he hadn't a worry in the world about his health.

In the winter of 1937–8, however, the strain of his work began to tell. In addition to the Sunday services at the Temple, he was preaching or lecturing in the provinces at least once or twice each week. Four afternoons a week he interviewed psychological patients. He was writing weekly articles for the *Methodist Times* and for the *Star*; there were church meetings to attend, and hospitals to visit. Then there were letters: at one time he was receiving on an average a hundred letters a day. Their contents were often so intimately personal that he felt obliged to keep them in confidence and deal with them entirely by himself. Letters would extend sometimes to thirty or forty pages, and once there was one of ninety; they would close with a remark such as, "I will do whatever you advise." The burden was dreadful: a life might depend on the answer he gave, and no answer might be given without long deliberation in untangling the skein of moral complexity. At the same time, he was incapable of putting off the responsibilities of brotherly love. "How well I remember," he wrote some years later, "between 2 and 3 one morning, putting down a letter, full of human agony, from a man in Australia who asked my advice, and finding, for the first time in my life, that I couldn't sleep for worrying about the troubles of the writer of the letter. I felt on a treadmill. I couldn't go on. But what could I do? After all, one cannot throw unopened letters in the fire. And if a soul in trouble cries for help one cannot say—'I can't help it, it's not my business . . .'"

Early in the New Year of 1938, in his letter to the church magazine, he outlines the activities of a typical day which had ended at midnight. But, he adds, "Please don't think there is any grumble in all this. To me this is the finest work in the world, and I think I must be the happiest man in London." Well, maybe; but happy men are generally less cerebral. Although the lot of the manual labourer has been sentimentalised, he does have the great advantage that, whatever he has to endure during working hours, he can walk away from when they are over: when the whistle blows, he drops his tools. Men whose labours are intellectual, on the other hand, do not discontinue them at the end of the day; priests are priests around the clock. In those days most Methodist ministers used to wear clerical collars and often black suits, day in day out, no matter what secular errand they might be on. And once when my brother and I were walking in the country in the August heat and had taken some ears of wheat and had stripped them to chew the grain, there came a great voice booming across the shimmering fields, "Ah, my boys, emulating the disciples, I see," and there in black amid the golden corn was Will Sangster, at the end of the lane in the depths of rustic England, wearing his dog-collar, on duty as always. Les didn't usually wear his clerical stuff when he was not preaching: he felt that it separated him from people. But in his mind he divested himself of nothing when he went out to play, on the few occasions that he played in those years. Now, in the early spring of 1938, however, he began to carry his burdens even into his sleep. When at the end of a sixteen-hour day he ought to have shed all its concerns and gone to sleep, he found that these were not to be disposed of. He remained restless, unable to relax. Sleep held off or came late, fitfully, and open to the invasion of busy dreams. In the morning he would get up exhausted in body and disturbed in spirit. Then one Sunday evening in the pulpit he felt a sudden pain in the back of the head, as if from some physical blow. He went to a physician, Lord Horder, who examined him and announced that emotional fatigue had so eroded his health that at the present rate of activity he would not live more than two years.

With this shocking verdict, Les eased up at once. He arranged to have the pulpit occupied by visitors through the month of March, and he gave up all his writing commitments and the out-of-town speaking. He expected to be back at work again by Easter. But the weeks went by: spring came, and early summer, and still recovery made only the most minutely perceptible approaches. In March, he went with my mother down to the Dial House. He had been ordered to rest, and rest he would, though inactivity is burdensome to a man who for years has jealously employed every minute of the day. He went on walks and read a little. But it was a fierce winter: the night closed in early round the house; the wind howled in the pines and gusts of rain beat against the windows. His head ached almost continuously; often for hours together there was nothing he particularly felt like doing, and he would sit in the living-room looking at my mother across the fire, nursing anxiety. The wireless would bring them the latest threats from Europe, and then Les would go to bed to keep his miserable sleepless vigil. The sounds that had been good the previous summer now added to his miseries—the noise of the sea birds, the wail of the fog-horn, the sea crashing against the cliffs at high tides—these became tiresome and repetitious, part of the boredom of the long insomniac night. In the early spring, Les went to Leeds to stay with the Appleyards, and they took him out on the Yorkshire moors to watch the birds, which were to become for him an absorbing interest. But he still could not sleep. He complained of the cuckoo. It is a slender hawk-like bird, as the bird books would say, a denizen of shady woodlands and thickety retreats. Its call consists simply of two descending notes separated by a two-tone interval. It is a migrant. Arriving in April, it earns some affection by being conventionally regarded as one of the signs of spring: regularly in that season the first sounding of its notes, for the bird itself is rarely seen, will rouse someone in the Home Counties to write to *The Times* and declare forthrightly, "Sir, This morning I heard the cuckoo. I am, etc." As the two notes are endlessly repeated, however, the bird's cry can become irksome. In Delius's piece, "On Hearing the First Cuckoo in Spring", which includes the bird's call, what is most manifest is not the freshness of the herald of spring but the oppressiveness of the notes. The bird is also associated with madness, the word cuckoo being slang for madman. Les felt the oppression and feared the madness. And when the wretched thing cuckooed all night and his hosts confessed that this was a thing undreamed of in their ornithology, Les feared he was having hallucinations, and his anxieties increased. It does sometimes do it though, he later learned.

It was a dreary time. Then in May, while he was staying in Cornwall for a couple of weeks, he began to sleep again. Sleep did not return regularly, but at least he knew the old joy, which can only be recognised as joy by those who have experienced its loss, of falling at night into oblivion and emerging from it refreshed. There were some good nights again for him, and he would wake to the spring countryside and once more relish life. Later in the spring he travelled round Ireland with his old friend, W. L. Northridge, who was Professor of Psychology at a college in Belfast. They went by car and it rained all the time. When they came to the Customs on their return, they leaned over and gathered up their clerical collars from the back seat, and the officer put his head in through the car window and said, "Well, Father, you wouldn't be carrying anything dutiable, now would you?" at which they grinned and drove on. They must have had a good time: Les returned not only with the Connemara tweed that the customs man wouldn't be thinking they would be carrying but also with a new song he had learned in a farmhouse somewhere, of which the chorus rang, "Oh Lucky Jim, how I envy him", lines with which he would break into song at moments of supreme happiness and release for years and years afterwards. They were not melodiously rendered (when he used to broadcast, the B.B.C. would ask him kindly to refrain from singing the hymns!), but it was good to hear them echoing in some rocky canyon when he was walking in his beloved Border country or, having picked his way across London, at last bursting out of the traffic into the open road to Dover, flooring the pedal of his Jaguar, "Oh Lucky Jim . . ."

At length, in June, Les saw his physician again, and in mid-July he was permitted to preach. It was only to be for one service, and he trembled at the prospect. He began his sermon by thanking his people for their prayers. He knew that even as he was speaking one minister friend had locked himself in his study for the duration of the whole service to be alone to pray for him. He said that none of his psychological principles and none of his spiritual beliefs had been weakened by the experience of his illness. He quoted lines from Browning:

> If I stoop
> Into a dark tremendous sea of cloud,
> It is but for a time; I press God's lamp
> Close to my breast; its splendour, soon or late,
> Will pierce the gloom: I shall emerge one day.

He was, he said, emerging. There was a large queue to greet him, and both the auditorium and the overflow hall were full. People had brought hordes of flowers to welcome him back—dozens of roses and scores and scores of carnations. My mother had to recruit help to carry them to the car, and she left the church in a profusion of blossom.

He was back again after five and a half months away. It was something. But he was not yet well; the single service depleted his strength more than he had expected. He was impatient to regain his old powers and take up his duties. Immediately after the Sunday services, however, he left London, and, with the Appleyards once again, travelled to Skye. He was back in London again to preach on the last two Sundays in July and to celebrate the sacrament of infant baptism. He was away in August and September and then he was back again for the first Sunday in October, which turned out, as we have seen, to be the day of thanksgiving for deliverance from war. Les had a very great deal to be thankful for on that day, if he still had much to fear: he was not perfectly fit again by any means, but, looking back on the depths from which he had emerged, he could be thankful.

He had been firmly instructed by Lord Horder to take it easy, though. And he was determined to try to do so. The church co-operated by retaining for one year an assistant minister, Dorothy Wilson, who for a number of months preached at either morning or evening service and lightened my father's load. He announced in the press that he would have to decline invitations. At the same time, to clear up misunderstanding, he let it be known that for much of his preaching and healing he did not accept fees. He quoted the announcement Joseph Parker had made about his charges:

> Preaching on behalf of the salaries of poor ministers: NOTHING.
>
> Preaching for ministers whose salaries are less than £100 a year: NOTHING.

Preaching at the opening of Chapels: Six volumes of standard literature.
Attending Tea-meetings: Fifty pounds.
Going to Bazaars: A hundred guineas.
Serving on committees: Two thousand pounds.

Les announced that he charged no fees for preaching away from the City Temple, but expected travelling expenses. No fees had ever been charged for psychological treatment and none would be, either by Les himself or by the five doctors in the Psychological Clinic. All profits on books sold in the church porch went to the church and in the case of his own books the royalties also. No fees had ever been received for broadcast services—this he particularly wanted to remark, since some people thought that the B.B.C. paid their ministers at the same rate as comedians.

Late in December 1938, seeking rest and health, Les left London with his elder son Dick for a cruise down the coast of South America. Once again, aboard the *Asturias*, his sleeping improved. He lay in a deck chair and watched the waves go by; he read, ate, played games, and revelled in the sunshine. It was good to be away from the telephone and letters and the need to make sermons; but, although he was strictly under doctor's orders, he felt no small twinge of conscience at leaving his people once again. By mid-February, however, he was back; and from then on he resumed his duties to an increasing extent, as his strength gradually flowed back, and he maintained them throughout the war, which by then was imminent.

7

All his life, from very early on until his last days, Les maintained an uncommonly broad credibility for the uncanny. "Uncanny" was a favourite word: his mind was drawn towards the ragged edges of the rational, where scientific explanations broke down to give scope to wonder. He believed—or rather, he refused to disbelieve—many things that most men—men who deal with practical things in a solid world—would automatically dismiss. He loved ghost stories; he more than half-believed in evil spirits; he was a member of the Society for Psychical Research; he followed with fascination the experiments of J. B. Rhine at Duke University; all kinds of extra-sensory perception intrigued him; he was interested in flying saucers; he attended séances. Anything inexplicable—pictures that changed places on the walls, knockings heard only when the young housemaid was in the house, vases that flew off their ledges and crashed down the stairs, gulls that flew wailing round the house where a woman lay dying, dogs with raised hackles barking at unseen presences—anything in that broad category was, one might say, his department.

I accompanied him to a séance once, along with some ministers and doctors, and we sat in a broken circle with the medium at one of the free ends. The medium's control, that is his opposite number on the other side of the great divide, was a little cockney boy called Mickey who had been knocked down and killed by a car while selling newspapers in Camden Town. He was a lively little kid—if one may use the word "lively" for someone who is dead; and when one of the ministers in the audience declared rather solemnly, in the words of John, that he must "try the spirits whether they are of God", Mickey said, "Nah, it ain't vat way over 'ere at al'." Mickey produced a number of spirits who had been acquainted with members of the audience before they had passed over. The brother of one of the ministers had been a soldier and he was anxious to describe the location of some trivial object he had left in a dresser drawer in his bedroom. We heard from one or two other war casualties who had left their bodies on a distant shore, but what they desired to communicate seemed equally insignificant, and they seemed eager to get back to wherever they had come from. We were sitting close together, elbows touching in absolute darkness. The spirits who had

President of the Methodist Conference, 1955.

Receiving John Wesley's Bible from the outgoing president of the Conference, the Rev. Crawford Walters.

With some American visitors at Wesley's Chapel, City Road, London.

Recording *The Will of God* for the John Milton Society in New York, 1954.

passed over ceased, for the moment, to communicate, and there was silence for a bit. Then was uttered what I think was the most chilling sound I ever heard: a mixture of a roar and a scream, long and drawn out. The minister next to me muttered, "Oh, my God," and although he was a Protestant began to cross himself excitedly. The room was suddenly filled with the scent of violets. Shortly, the woman to my left, who was supposed to be part of the audience, announced that she had inadvertently gone into a trance: the roar we had heard, she explained, was that of a panther, which had been the pet of a young woman and had been killed along with her when she had been escaping from a harem. This had happened in the Middle East four thousand years ago, we were informed. The beast seemed still in good voice. "If you lean forward," said the woman who had stage-managed this latest phenomenon, "you can touch him." "I can feel him," someone cried out in the darkness across the room. "He went by under my hands."

It was pretty nice after that to come out of the house into an ordinary suburban street with the stucco houses and pink tiled roofs in the usual watery London sunshine. Les said, "Isn't that interesting! I'm not sure about that panther though—four thousand years is a bit of a long time." But, as was his way with these performances, he suspended his disbelief. Once in a séance he attended, John Wesley turned up; and once, in a hushed whisper, a spirit said of the earth-planet, "This is the only one that has gone wrong!" Les attended séances repeatedly with a war widow, and her dead husband would identify himself by offering his own familiar endearments. Once he spoke to them about a dog his mother had purchased of which the widow had not yet been informed; he proved to be correct, and they were impressed.

One of the most extraordinary communications he received from the dead came to him in the mid-fifties when he was suffering from a protracted period of ill health. It was a typescript of automatic writing that had been executed by Geraldine Cummins, having been dictated by a spirit on the other side, called Astor. This spirit was concerned about Les's health. With him, also concerned, were two other souls who had had an intimate connection with my father: one was Andrew, his father, and the other was Maxwell Telling, the physician who had helped him with his early psychological work in Leeds. The spirit of Dr. Telling came with a physical diagnosis for Les and medical recommendations. But Astor could see "from Andrew" that one factor in his ill-health was the racial memory: the unconscious mind, Astor points out (knowing, presumably, his Jung), is a reservoir of memories going back beyond the life of the individual into that of the race. In Les's unconscious there was still the deep memory of the insecurity with which his forefathers in the Border country had lived, menaced by cattle thieves, clan hostilities, and threats from English marauders in the south. These memories made their disturbances and brought present anxieties. I don't know whether these curious tidings from Astor brought any relief. But it must have been something for Les to think about—that battles of broadswords and targes and the terrors of herdsmen beleaguered in their towers still raged in his blood, while he watched the rain on a dull Saturday afternoon in Finchley, for instance, and wondered whether to go out, or while he was stuck in the traffic, waiting dully for the lights to go green.

Without always knowing what to believe in, Les knew there were more forces at work in the universe than science had codified or philosophy dreamed of. He himself not infrequently received messages by means that were inexplicable. There were things that he learned by a sixth sense, and on checking up on them he would find that he was correct. One such incident occurred when he was staying with an old friend and preaching in a small country place. A man with his wife sat near the window at the back of the chapel. As Les looked at him from the pulpit, he *knew* that the man was passing through a financial crisis and he accordingly modified one or two parts of the sermon to emphasise that all kinds of disaster were proper occasion for prayers asking God's help. Afterwards his host remarked how apt the sermon had been, since one of the church members had recently gone bankrupt. "I know," said Les and proceeded to describe the man and the place in the chapel where he had been sitting. This was not an unusual occurrence. Once he similarly identified from the pulpit a woman who was unhappily married, and once a couple who had lost a child. He was gifted with some unusual kind of perception which he could not explain.

One of the most extraordinary of the inexplicable forces in the world was the power of healing. For years, pressing along more or less alone with his healing practices, Les had advocated co-operation between minister, psychologist, and doctor. Each alone offered an incomplete contribution to healing, he thought: the doctor's method was material; the minister's spiritual; the psychologist's procedures were analytical and did not encompass synthesis—he did not, in Leslie's phrase, know how to put Humpty Dumpty together again. There was of course some opposition to Les's aspiration of having these three kinds of professionals work together: people felt that healing was exclusively the domain of the medical profession and that there were hazards in introducing unqualified practitioners. Ministers, it was claimed, should stick to their lasts (there is always a proverb to hand to support the conservative position in a debate!). Les responded with verve to such arguments: doctors, he had discovered, had very little training in psychology, if any at all, and they were not qualified to treat psychological patients. That ministers should shun the healing business, he declared, made nonsense of the healing miracles of Jesus or indeed the enjoinment of James, the Lord's brother, "Is any sick among you? Let him call for the elders of the church; and let them pray over him . . . and the prayer of faith shall save the sick."

At last, at the Methodist Conference in the summer of 1936, the report of the Committee on Spiritual Healing that Les had chaired was moved and carried. His speech showed him at his most persuasive: the first business of the Christian church, he said, was to bring people into a vital relationship with God, and against that condition the question of health dwindled into relative unimportance. There were, however, terrific energies passing through the personality of which no one knew the strength or the limits: under certain conditions it seemed that no disease might withstand them. On the other hand, nothing was known about their operation: one man gets better and another does not, and no one knows why in either case. The report advocated more prayer for the sick in public worship, more training in psychology for students during their college course, and the setting up of clinics of doctors and researchers so that co-operation could be effected in efforts to heal the sick. Les petitioned that the committee be re-appointed for further research, and the Conference adopted the report with enthusiasm. It would have been difficult to resist the appeal: one newspaper reporter notes in his column that Les looked young and healthy and handsome as he delivered the report. He was widely loved by his colleagues. He had just been appointed to the highest preaching position in the country, and he had asked that he might keep his place among them. He felt that the adoption of this committee report was a milestone in the progress of spiritual healing.

Some of the extraordinary manifestations of the healing powers that psychological treatment could make available have been observed above: in Leeds in the bare rooms opposite the huge chapel my father had interviewed many people in distress, physical and mental, and had sent them away cured. When he came to London, he organised his healing on a larger scale by obtaining the services of professional doctors and psychotherapists who were also Christians—or rather, to set their qualifications in the order in which Les ranked them, they were devoted Christians who were also professional men and women. Over the twenty-four years of his ministry in London the number of people who came to the clinic to be unburdened of their manifold sufferings is not to be computed.

But there were stranger methods of healing, less explicable than psychology, even in the earliest days of that science. In Leeds, Les had begun to incorporate into the evening services the practice of offering intercession for individual sufferers. The lights in the church would be lowered. There would be general prayers followed by a long period of silence. Then Les would ask for the prayers of the congregation for certain individuals. Sometimes, with the permission of relatives, these would be named; sometimes they were merely designated as a father, a mother, a young woman, a man in his forties, a child of seven, as it might be. Their condition would be described, generally in unsensational words, though sometimes the pathological terms would be used. Les would be most concerned to make an imaginative picture of the person for whom the intercession was requested. Then he would say something like this: "Don't think that you are trying to persuade a reluctant God to heal the sufferer. And don't pray that he may become better, for that is to put his cure in the future. But imagine you are bringing him into the presence of Christ and that Christ is laying his hands on him. Believe that at this moment His healing power is working." Then there would be silence. Then, "Don't let your mind wander. Let it be concentrated, watching, helping Christ in the healing." Then more silence. There might be as many as four or five persons individually named or described for whom prayers were requested.

In addition to the Sunday night intercessions there were meetings of smaller prayer circles whose members would meet every day in the week to pray for individual sufferers. For some people, prayers have been offered for months and occasionally for years together by people who believed in this method of helping their fellow men and were willing to devote their time and indeed their strength to this work. Prayers were asked mostly for those whose troubles lay beyond the scope of conventional healing or those for whom this had proved no use. Sometimes miracles occurred; sometimes nothing. Les didn't pretend to know much about the relationship between cause and effect in this whole phenomenon. After twenty-five years of practising such intercession, experiencing mostly "the blank misgivings of a creature moving about in worlds not realised", he came, however, to one or two conclusions, vague enough and tenuous, as to the rules by which it operated. It seemed, for instance, that prayer worked better for children than for adults, perhaps because their minds were not closed by preconception against what is possible and impossible. Prayers for people known and loved by the congregation seemed more effective than those for people who were strangers to it. Those making the intercession must concentrate and they must believe that they were co-operating in recovery.

Les was more unwilling than most of his critics to claim that prayer cured physical illnesses. He was at pains to distinguish the mere alleviation of symptoms from a thorough cure; unlike the Christian Scientists, he was scrupulous in giving their due to physical methods of healing. In some cases he would claim for prayers only that they might have established in a patient a receptive state of mind in which the conventional medical treatment would be more effective, as an oxygen tent put up over the patient creates an atmosphere conducive to healing. Hundreds of people told him that the turning-point in their sicknesses had been that at which they determined to become well or had become strangely but firmly convinced that they would recover.

He was antagonistic to mass meetings in which sufferers were invited to come up and be touched by the practitioner and encouraged to expect miraculous restoration. There is a long and dismal record of such meetings which have repeatedly brought cruel disillusionment. The blind and the crippled are led forward to the platform and, still blind and crippled, they are led back again. One feature of these situations that is insidiously encouraging is that a temporary alleviation of the symptoms of sickness is often effected: the crippled find use of their limbs for an hour or two and throw away their crutches. Then they wake in the morning to the old paralysis, sometimes worse than before, and they despair. Les came to hear of many occasions when so-called faith-healers, most often in innocent stupidity rather than calculated fraud, performed what in effect were cruel theatrical tricks. And he joined other leaders of the church in warning people against them.

He was in many respects at odds too with Christian Science. He objected to the theory that denied the reality of matter—the old Gnostic heresy that Christianity had hunted down in the Middle Ages. If matter did not exist, Les felt, then the Incarnation was a delusion. But he objected even more strongly to the well-known Christian Science principle that physical methods of healing were to be avoided. His own great enthusiasm for non-physical healing never diminished his tremendous respect for doctors and their performance; if, as not infrequently happened, his own success with a patient upstaged conventional medical procedures, he never weighed one kind of therapy against another; he had wanted to be a doctor *and* a minister; and now that he was a minister, he wanted to co-operate with the doctors. The Christian Science policy of keeping the doctor away he denounced as wildly impractical, illogical, and wrong.

He recognised the great merits of the faith—the emphasis on the reality of the spiritual world, for instance, or on the power of mind over body. But, characteristically enough, he reduced the matter of fighting disease with physical methods to an illustration: if Mrs. Baker Eddy, the founder of Christian Science, had lived in a Russian village repeatedly beset by wolves, would she not have resorted to physical methods to resist their onslaught? She would. Why then should not the attack of bacteria upon the body be withstood with whatever methods may be available? It should. He used to get quite exercised about what he considered a narrow superstitious unwillingness to consult the doctor in the case of sickness. Adults, perhaps, might be let alone to suffer for their own purblindness (though he took a little wicked delight in reporting that Mrs. Eddy, when afflicted with raging toothache in the middle of the night, abandoned her metaphysics and sent her students about the neighbourhood to scare up a dentist); but the prospect of children being made to suffer for the beliefs of parents angered him. Les was tremendously fond of children; they feature in all his writings as the innocents, and no evil, in his scale of values, was so vicious as that which injured them in mind or body.

Like so many miscellaneous faith-healers who call upon the lame to throw away their crutches, Mrs. Eddy claimed some sensational cures. Looking back over thirty-five years of his own practice of asking for intercession, Les made the sanguine estimate that in less than 5 per cent of cases had prayers been followed by any marked improvement in the patient. And, he adds, "even then . . . it was open to any critic to say of the patient, 'He would have recovered in any case', and nothing we could say could ever prove that prayer was the decisive factor, although in some cases it seemed the most likely therapeutic factor." The small percentage showing improvement, however, did not include the people who responded not by physical recovery but by the strengthening of their attitudes towards illness. Not infrequently a patient might say, "I felt that the illness didn't matter any more."

Five per cent seems a conservative estimate in any case. Some of the apparent results of the devout prayers of two thousand people must have more than compensated for the larger proportion of occasions when there were apparently no results at all. And what was often so impressive was that healing began within the very hour, or less, in which prayers had been raised. One Sunday evening just as he was about to leave the house for the City Temple, Les picked up the phone to hear the voice of a young woman whom he had known for many years. She was distraught with grief; her seven-year-old son had developed a severe case of polio and lay pitifully sick with a high fever. The case was so desperate, she said, that the doctor who had just left the house had declared that the boy would be dead before morning. Would the people at the City Temple pray for her son, she asked. Les told her that at seven o'clock that evening, the regular time for intercession, prayers would be offered. He asked his people, "Don't let your mind wander. Watch and pray. Believe in recovery." When he got home he telephoned the mother to tell her what had been done and was astonished to hear her say, "Dr. Weatherhead, he is sitting up playing with his toys." He had sat up first at seven-thirty and asked for something to eat. His temperature had dropped incredibly. Prayers for the boy were continued both in the church and in the prayer circles, and in two months Les learned that the boy had surprised the doctors by learning to walk again more quickly than they had expected and by the improvement in his muscular tone.

Over the years there were other bright miracles to lend some radiance to the routines of intercessory prayer. A nurse in a London hospital had been sleepless for some days and nights and her temperature was up to 104. She didn't know that prayers were being offered for her, but, on the same Sunday evening her temperature fell to normal and she slept without a drug. A young airman, who had been told his disease was incurable, had been able to return to work after prayers had been offered for him. A young expectant mother, paralysed in both legs, who had been told she would be unable to deliver her child, had walked within a few hours of the time that people had prayed for her and had later successfully given birth. One woman wrote to Les (with an unconscious pun), "Can you tell me why it is that, being blind and having been given up as hopelessly blind, at 7.30 last Sunday evening it was just as though a blind went up and I saw and have continued to see?"

The record of one instance of the effect of prayers is carefully documented with the comments of the doctors who had charge of the case. (There were other such records, but they were destroyed in the bombing of the church.) It was a case in which the healing process did not strikingly coincide with the time at which prayers were offered. A boy of four and a half was hospitalised with nephritis. The church congregation began to pray for him and continued each Sunday for six or seven weeks. Nevertheless, after a few weeks in hospital the boy was sent home with "permanent" damage to the kidneys and the expectation that he would die. His body began to swell; gradually the swellings rose up to the level of the heart, and he went into a coma. Both the doctor and the nurse attending him said now that this was the end—he would last only a few hours. The father sat up that night to watch him, and, to his amazement, in the small hours of the morning the child regained consciousness, sat up and asked for his mother. Shortly he became perfectly well. Tests revealed that his condition was normal. The doctor in charge of the case wrote that he was "quite sure" that "prayer played the biggest part in his cure. He was suffering from a severe sub-acute nephritis, and in the opinion of all the hospital physicians his prognosis was of the very poorest." Another doctor wrote to say:

> I have attended a number of similar cases and never saw one recover, and owing to this and the absence of any sign of improvement, I could not give any favourable prognosis to the parents which would allay their deep anxiety. One day which I shall always remember, the father asked me if there was any hope at all for his recovery, and I had to reply that, as far as I could see, there was definitely no hope whatever.
>
> Within about a week after this, David's condition had completely and suddenly changed for the better, and in a very short time he was practically normal which was most extraordinary to me.

On September 1st, 1939, when Hitler sent the *wehrmacht* into Poland, we were all down at the Dial House at the end of what proved to be the last of the good family summers. There had been days of swimming and canoeing in the bay and walking over the cliffs and watching the ships, and nights with the owls hooting about the house. When war became imminent Les decided he ought to be in London with his people, but he left the rest of the family at the coast. We took pickaxes and shovels and dug a deep trench in the chalk among the pine trees and stocked it with a ladder and a bench, a couple of flashlights and the gas masks, a tin of Huntley and Palmer's breakfast biscuits and a chamber-pot. We anticipated, obviously, sieges of only limited duration. Sunday, September 3rd, was a glorious cloudless morning. We came in from the garden to hear Neville Chamberlain announce over the radio that we were at war. "You can imagine what a bitter blow this is to me," he declared, and I remember feeling what was later expressed by a sailor in Noel Coward's *In Which We Serve*, "It ain't no bank 'oliday for us neither!" That night the sirens sounded and we scampered down through the garden into the trench where we sat huddled, elbows on knees, and listened tensely for the sounds that would announce our annihilation. "Whatzat!" "There's a bomb!" we gasped at the sound of a low sort of rumble. "My stomach," said my mother. In the morning the milkman found us, stiff and damp, still in the earth, and told us it had been a false alarm: a formation of geese had been mistaken for raiders crossing the coast.

Meanwhile my father had installed himself in a hotel in Kensington and was preparing an address to meet a situation which surely in those last days no one could fully believe would issue in a new deliverance. When it became known that Chamberlain was to address the nation on the Sunday, my father had a radio rigged up in the pulpit, and thus the church congregation was able to listen to the fateful message. A little later they heard the siren for the first time, that wailing banshee that in a few months was to become such a regular feature of the constricted lives of Londoners, who were to hear it with wearisome repetition day after day for weeks and months together in those grey years—at dusk, as they put up the black-out frames over the windows, or as they scuttled into the Underground to pick their way to the trains, or as they stood in cold drizzle waiting for a bus while the early searchlights swung across the clouds. The miserable wail was to become something usual—like rationing, and shortages, and the untimely death of friends.

But that was later. Now it sounded in church on that bright September Sunday morning; and Les, standing in the pulpit, was moved as he watched the congregation, quiet and resigned, trooping to the basement. It was the first Sunday of his fourth year at the church. He was shortly to turn down a chaplaincy in the Air Force, and he was to stand by here in London and see it all through with his people.

After lunch that Sunday he went for a walk in Kensington Gardens with the owner of the hotel where he was staying. They came across a dead pigeon on the path and, to the surprise of his friend, Les took out his penknife and amputated the bird's foot. This, as he explained, was for his younger son, who had an assortment of feet taken from dead birds and placed in little glass tubes to satisfy his collector's instincts. It was typical of Les, in the hour when empires dissolved and heaven was falling, to have in mind such needs as dead birds' feet for boys. Never, whatever the circumstances, did he go oblivious of the simple needs of people. It was, I believe, part of his genius that, with the ability to speak to thousands, he could speak also to single individuals; that he could leave a broadcast service, for example, in which his voice had been relayed over half the world, and go to his car, enquire intelligently about the family of the car-park attendant, give him a can of marmalade from his overcoat pocket, and tell him to treat his wife's rheumatism with an electric heater. He could do this perfectly naturally without being patronising, because individuals mattered to him. Later in the war Americans sent him CARE parcels, and often enough the janitor of his office building o the car-park man, or the man who sold the *Evening Standard* at Holborn Viaduct would feel pressed into his hand a tin of butter or Canadian bacon or the pink rubbery processed meat or the yellow rubbery processed cheese that were so delectable in those hard days when, as Alice Head remarked, you were lucky to be able to lunch in your office off half a seagull and a vitamin tablet! Those whom Les encountered with these goodies must have looked forward to the appearance of the man in the Anthony Eden hat and the dark buttoned-up overcoat with the bulging pockets who would ask after their wives and children.

He could be natural with anybody. Even with children, who inhabit for the most part their extraordinarily solipsistic universes, he could be perfectly at home. He liked them. He could be stern if they menaced his peace, and occasionally he expressed sympathy with the disciples on the occasion that they had chased the children away from Jesus just before the Sermon on the Mount. But he not only liked them—he also held them somewhat in awe: they were for him symbols of innocence, if they were small enough: in his imagination they belonged in unblemished gardens in sunlight more radiant than the world actually knows. They in turn were drawn to him to share pleasures and confide concerns. They came to him after Sunday services and he bent in his cassock to hear their whispered secrets. Little girls occasionally offered him marriage. He knew how to act with children, and with teenagers who came to him with their teenage heartaches and their melodramatic gloom. He knew about them all and what they were doing, remembering perhaps his own childhood and how he had hungered for an adult friend. Often he would write letters to children if they were sick or especially when they had examinations coming up—ordeals he himself had dreaded as a boy. Alice Head had given him a magnificent fox-fur rug with the fox-tails still dangling from it at intervals, and as these dropped off from time to time, Les would stuff them in an envelope and send them to some child who needed encouragement, with a note saying, "Keep your tail up!' So to a boy missing his father at the outbreak of war he sent the pigeon's foot with a note of cheer and encouragement.

In the first few months of the war he spoke repeatedly of the things in life that war could not injure, such as the beauty of art and music and literature, the friendship of men and women and the charm of small children. It need hardly be added that prominent among the things of invulnerable and permanent good were the things of nature. Following a life-long habit when beset with troubles, Les looked away to the mountains or, if they were not in range, he re-created them in his memory and bent his spirit towards them. They symbolised the things of God. They endured. And the smaller items in nature also reinforced his sense of God's presence and purposes. "If I give you a rose," he quoted from Tertullian, "you won't doubt God any more." For himself he had only to hear a blackbird singing in the rain to restore his belief in miracles.

He found and gave comfort in comparing the present with other dark passages in history or with the suffering of others which had blossomed into intensified faith. He described the scene in the upper room in Jerusalem after the crucifixion when the small band of disciples might well have judged that all was lost with the loss of their leader. He recalled the sufferings of St. Paul: "Five times received I forty stripes save one, thrice was I beaten with rods, once was I stoned, thrice I suffered shipwreck, a night and a day have I been in the deep, in journeyings often, in perils of rivers, in perils of robbers . . ." And by setting present distress into a broader context he provided some relief from fear.

His own faith, fortified perhaps to meet the evil events and growing stronger even as the news grew worse, became more comprehensive than before and increasingly outspoken. He believed in what he called a "*worthy denouement*" for every human life: "'not one life shall be destroyed'", he quoted from Tennyson.

> No, not one life, in any of the ravished countries or in the oceans or Norway's dark fiords. No brave airman, British or German, Grecian or Italian, falling out of the sky can fall out of God's hands. No little Abyssinian child, scarred for life by the mustard gas which Italy used to bring "progress" and "civilization" to a backward people in the "dark Continent", is one moment forgotten by God. No Christian pastor, tortured and sexually assaulted by the perverts who run German concentration camps will have suffered in vain.

That was something indeed to sustain people in the bad days. He felt that it was right to resist Hitler, and that feeling was strengthened as German planes bombed English houses and the German army conquered and enslaved one nation after another, bringing terror and cruelty to preside over communities of peace-loving people. But he was anxious also to resist bitterness—to scotch at the beginning of things the spirit that would ruin whatever reconstruction might come at the end.

Christmas came, even in wartime, even in the years of bombardment. We looked for gifts for each other and after the usual seasonal secrecy and smuggling produced them on Christmas morning before church. Gifts were not Les's department. Sometime in February or March he might say to one of us, "Where did you get that from?" and we would reply, "You gave it me for Christmas." "Ah," he would say, knowing that we knew that my mother saw to these things. "Remember," he used to say each Christmas (until, thoroughly familiar with the apothegm, we would say it for him), "remember, what counts is not the gift of the lover but the love of the giver", a philosophy predicated, no doubt, on the fact that, except for buying shoes, he hated shopping. He was generous enough and would take great pleasure in giving—the half-crown pressed into the palm when shaking hands with us as schoolboys, later the five-pound note—the old crinkly kind that would buy things. But except, curiously, to buy us shoes he hated going to shops, and he left all that to my mother. Sometimes at Christmas she would express her anxieties about protocols. "Oh, Les, we never gave anything to the Bradleys," she would say, and he would run off his formula about the love of the giver and say it didn't matter. He wasn't too great on gifts received, either, being unable to remember who gave him what. Once, ferreting through his closet for something, he suddenly turned to me and said, "Let me give you a shirt," and drew out and pressed upon me the golf shirt that *I* had given *him* not twelve months earlier for his birthday. It was a very fine shirt: I remembered it had cost twelve dollars. "I can't wear that kind of thing," he said. "Don't show it in Ohio, though; Dick may have given it to me!"

After church on Christmas morning my mother would scout around to find if there were any lonely people about who were going to eat by themselves. She generally rounded up four or five completely miscellaneous clients, mostly oldish, not necessarily poor, and ordered them up to the house for the day. I sometimes wondered if they really wanted to come, but she made no question of it. "It's Christmas," she would say, settling the matter. She coralled them into the car, sending us children home by tube. There was first the enormous meal and then there were the games. These might be intellectual and sedentary to be performed with a pencil, but they weren't necessarily like that. And it is some measure of the strength of my mother's will that, after the great feed, when one wanted only to roll on the settee, she had everybody down kneeling on cushions to play the balloon game—people hundreds of years old, it seemed, with arthritis and indigestion and no doubt heart conditions, batting balloons across the room. The very thought of it now, thirty years later, is enough to arrest the gastric juices! One could only hope that the balloons would hit the holly or that the King's speech would come on the radio and my mother sound the retreat. Later, there was cake and tea and prizes for everybody. And then the little old men and ladies went home on the tube to their individual peace. The day following, my mother would be up early to catch the postman and give him his Christmas-box.

In the spring of 1940 the City Temple celebrated its tercentenary: three hundred years before, Thomas Goodwin had come back to London to resume his Christian ministry. There was a week of services and meetings which included a mid-week public thanksgiving meeting, chaired by the Lord Chancellor, Lord Simon, and addressed by the Dean of St. Paul's; Dr. Sidney Berry, Chairman of the Congregational Union; Dr. James Black; Dr. Robert Bond, Moderator of the Free Church Council; Dr. Maude Royden, previously a minister of the Temple; and my father.

One feature of the tercentenary occasion was the publication of the history of the church by Albert Clare, which recorded the characteristics of all the ministers of the church. To some degree these had all been popular preachers, and in a foreword to the book my father discusses popularity; he complains with some gusto that in a preacher it is frequently condemned as a mark of shallowness, or it is thought to have been obtained by flattering the prejudices of those present, or by trouncing for their faults those absent. Why, asks Les, is it so culpable in a man to have people crowd to hear him preach?

> If I quote Thomas Aquinas, and Bergson, and Otto, and Dibelius, someone is sure to think it is a clever sermon, and intellectual preaching . . . If, having paid studious regard to all I can learn from such masters . . . I take pains to write simply and illustrate in a homely way and clear away doubts and beat out a track so that a boy of fifteen or a charwoman or a miner or a university professor can equally easily find their way to God, then my effort is damned by some highbrow in his first year at a theological college by the label "popular".

Les must have felt pretty strongly about this matter at the time. And certainly it always needs saying: clarity in speaking calls for intelligence and hard work, and scholarliness should not be associated with the half-articulate and obscure. But Les always protested he was not a scholar—he thought that scholarship was a field of endeavour separate from his own. It was one he admired tremendously—never, for example, under any circumstances did he omit to include people's degrees after their name. And in speaking of them he would often name the university where the degrees were earned. It is true that there were many intellectual pursuits about which he was not well informed: in science and philosophy and the arts he had not kept up. He knew little about Whitehead or Russell, Wittgenstein or even Sartre, except that he was a cousin, incredibly enough, of Albert Schweitzer. I don't think he had ventured much, either, into those dull hinterlands of learning where behavioural scientists make graphs and tabulations to render human nature in statistics. He wasn't particularly knowledgeable in politics: he believed a man should do a day's work and be proud of it, and he was against taxes. And once when James Callaghan, as chancellor, was photographed in his characteristic posture, Les appeared in *The Times* with a brief letter: "Dear Sir, It was a pleasure to see Mr. Callaghan with his hand in his *own* pocket."

When it came to theology, Les usually knew his stuff; in this field he was perfectly justified in implying that he had paid "studious regard" to the masters. He had, as we have seen, his own daring and original ideas, but he had kept up with current work. To decide whether or not he was a scholar in these matters would be to cavil over the ninth part of a hair: he might easily have made his name as one. If his work doesn't read like scholarship it is in part because the style is so beautifully limpid. A friend once said that if he called a discourse, "An Enquiry into the Conditions of Mycenean Civilisation in the Heroic Epoch, with Special Reference to the Economic and Domestic Function of Women, before and after the Conjectural Date of the Argive Expedition against Troy", he would be a scholar; if he called it "How Helen did the Housekeeping", he would not.

In psychology his credentials were equally secure: he did in fact earn a London Ph.D. in this field by writing a thesis that later appeared as the book, *Psychology, Religion, and Healing*. He never thought of himself as an academic, however; and when he went to Edinburgh for the conferring of an honorary D.D., he remarked, "I hope they don't ask me any questions!" But for all his modesty, his speaking and writing, no matter how widespread its popularity, had a solid backing of learning and devoted preparation by which it was differentiated from the glib and flashy orations of popular evangelists.

In the matter of the arts, his tastes, I used to think, were sometimes a bit curious and he was a bit behind the times. Lines of verse he saw fit to quote in his books could occasionally be quite without merit:

> Unto all peoples of the earth,
> A little child brought light;
> And never in the darkest place
> Can it be utter night—

He was liable to quote such pieces, however undistinguished, if they said roughly the right sort of thing. In music he liked melodious work, the nineteenth-century Germans particularly. He could play the piano himself with terrific panache: he played without music and could reproduce melodies and make up a grand sweeping orchestration in the bass—a talent that was useful to him in mess hall and church hall alike and in any informal gathering where anyone wanted a tune. And once in the old hotel at Stalheim, at the head of a wild windswept valley in western Norway, while he was relaxing with his pipe after dinner, a couple of teenagers came up to him and asked in broken English, "Shall you play on the piano so shall we dance?" and Les, rising to the occasion, began to hammer out tunes, albeit tunes twenty years old, while all the hotel guests, Norwegian and otherwise, came to the dance. He didn't go much for modern music though, on the whole; he disliked the cacophony of it. He disliked modern poetry for its obscurity. I remember once his arguing about Dylan Thomas with a Pole who was sculpting his head. "Well," said Les, closing the argument with aggressiveness in modesty's clothing, "*I* don't understand him!" He was like that with pictures too: he thought that they ought to be *of* something and that that something ought to be clear. Sometimes in the summer, coming home from the City, we would drive over Hampstead Heath, and I would stop the car and ask him if he would care to look at the pictures which used to be set up and shown for a week or so along the pavement near the Whitestone Pond. We would park and walk down the length of the exhibition: Les would grunt a bit over an abstract or a bit of surrealism: then, "What's *that* supposed to be of?" he might demand to be told, or, applying his most familiar aesthetic criterion, "That's not *my* idea of a horse!" Often, sitting cross-legged against the wall, were the painters themselves: they had black berets and two-day beards and it seemed pretty obvious that they were the unsmiling creators of these bizarre new visions of the world, and the owners of this or that idea of a horse. I would nudge Les and try to get him to shut up. "What?" he would say, "I could do better myself!" Part of this reaction, of course, was his characteristic suspicion of pretentiousness: he bridled when canvases which posed as art appeared to him as mere adventitious daubing.

His mind was not closed, however. When Epstein came out with his "Consummatum Est" in 1937 and fluttered the conventional cultural world, he had no stouter defender than Les. It was a two-and-a-half-ton pink alabaster Christ, recumbent, eyes closed, hands open to reveal the wounded palms, and like other earlier work of Epstein's it aroused indignation among the bourgeoisie who saw in it a challenge to Christian principles and morality. "Give credit to the alabaster Christ," wrote Les. And then again later when "Genesis" appeared in a London park, he came flying to the defence: Epstein, he argued, knew how to make conventional sculptures; he knew what he was doing in these momentous departures, and he ought to be trusted. Les may have recalled in these exchanges how often he himself had ruffled the contentment of conventional acceptance. But in general, in the arts, he liked the older forms: he liked very little of the stuff that expressed the racked experiences of the modern consciousness: he preferred Thorvaldsen's Christ to Epstein's—certainly he preferred it to any of Graham Sutherland's. He wanted nobility in literature, music, and painting; and, of course, he wanted peace: the arts should provide a golden world and a space in which the oppressed spirit could withdraw and lay by its troubles.

In his address at the City Temple tercentenary celebrations, Lord Simon declared that it was British liberty that Hitler was trying to overthrow. Within a few weeks of this less than sensational pronouncement, Hitler, having overrun Norway, Denmark, France, and the Low Countries, stepped up his assault against England, and in early July the Battle of Britain was joined.

Some time during the first winter of the war, the south coast of England had been closed to civilians—except of course those who lived there. And so, when corresponding with a headmistress of a girls' school in Buckinghamshire, where he had distributed the prizes, my father had casually enquired whether there were any quarters she knew of in that area that he could rent for a week or two in the summer for the family. The headmistress, Norah Laycock, kindly offered him the lodge of her own school, which was to be unoccupied for part of the summer. The school was situated in the Chiltern Hills thirty miles out of London. It was experimental, and the girls, instead of playing games, tended the animals of the school farm. It was a fine estate in good country. There were woods and fields and a walled-in garden where the sun saturated the old brick and ripened the peaches and damsons trained against it. My father set up a deck chair in a warm corner of this garden and used to read and write to the sound of linnets twittering in the fruit trees and the hum of bees. We were all here among the green fields when the Battle started.

The bombing began on the ports of the south coast and on airfields south of London. Then one night we woke up and looked out to see a whole quadrant of the sky red over the horizon and above it a glowing haze. There were telephone calls early next morning, and my father had

With Frank Salisbury who painted his portrait.

Welcoming Queen Elizabeth the Queen Mother at the dedication of the new City Temple, 1958.

The new City Temple.

to drive into London. As we drew near, the air began to smell of burning, and then came the sickening smell of burned timber that has been doused with water. Occasionally there was dust from masonry that suddenly broke loose and fell, and once there was the roar of a burning jet of gas from broken mains. Some roads were strewn from side to side with broken glass which crunched under the tyres. People were sweeping it off the pavements like so much snow. A number of roads were blocked; over others ran fire hoses as the auxiliary brigades still shot water into the black husks of burned buildings. There were little wooden ramps so that you could drive over the hoses without crushing them. Never, I think, did my father so identify with the city and its seven million souls as when he saw it in its distress.

That night, in which six hundred planes had been over central London, was the beginning of the terrible onslaught on the City which was to extend without relief for fifty-seven nights. Goering had ordered it to be undertaken in the expectation that the resultant confusion and grief would bring Londoners to the pitch of revolt and England to its knees. Every night from September to November an average of two hundred German raiders attacked this city. Every night ten to twenty thousand people were rendered homeless; hospitals were blasted open and wrecked, oil tanks and wharves were set on fire, ships, warehouses, towers, churches, public buildings burned, and even the wooden blocks of the roads; burning barges drifted down the river, drains were smashed, and light-cables severed. Hundreds of thousands of people spent the long hideous nights in shelters, sometimes with little or no sanitation. Never had a civilian population been besieged for so long with such weapons. Even far out in the suburbs the nights were loud and dangerous with gunfire; and in the daytime little ladies walking their small dogs in municipal parks would pick up jagged chunks of shrapnel and carry them home to lay on the mantelpiece by the china ornaments and the coronation saucer with the picture of the king and queen. The war came home to all, and everywhere among those killed, injured or otherwise outraged were my father's people.

He opened his fifth year at the City Temple on the last Sunday in September after the most terrible week in London's history. The church was, as yet, almost intact: only a bomb which had hit the *Evening Standard* offices across the street had blown out some windows. It was eerie to be in the City on a Sunday morning among the open wounds of the action of the previous night, and it was dangerous. One was not to know, of course, when the next raid would come, if a lone raider escaped detection at the coast. In those days, bombs might come, as they say, from a clear blue sky; and then there were bombs with delayed action fuses, deposited undetected the night before, which would blow up without any warning at all. The evening services had been cancelled so that people would be at home when the raids began and the shrapnel came rattling down on the asphalt. But on Sunday morning there were substantial gatherings at the church: the galleries had been closed, but the ground floor was filled.

Les spoke in those days of love and faith: it was the same message of the last twenty centuries but now applied with special insistence: "God can do much with those who have passed through the fires of suffering . . . I think God can do nothing with those whose hearts are full of bitterness or who allow themselves, by the spectacle of temporary devilry, to be driven into the darkness of unexpectant despair." He spoke too of the church and its role in those hard times. He reminded the people repeatedly that it was "a house not made with hands". And it was as well to recall its spiritual body, since the physical fabric was shortly to be destroyed.

In the meantime his people here and there were afflicted, and he spent an increasing number of hours picking his way across the city, often through its ruined areas, to visit homes and hospitals. He described the experience of a single afternoon when he made contact with a variety of the kinds of tragedy war had staged. First he visited a girl who, in Les's own words,

> in one night, had all her family killed save a sister who was a nurse on night duty and who came home in the morning to find her home a mere heap of rubble. The girl I visited was trapped by heavy masonry falling on her feet. She could move her arms, however, and had the presence of mind to wriggle until she could reach an alarm clock, which she manipulated until the bell rang. The rescue squad heard the sound and dug down until they set her free. My next visit was to a home where a family of ten is reduced to living in a single room of small dimensions. The other rooms have been made uninhabitable by bombs and fire. My next visit was to a small shop, which I am afraid will never do any business again. With it goes the means of livelihood of two elderly ladies.

Such dreadful fates became the regular pattern of those days: death and destruction moved off from the headlines to remoter columns. The carnage in all its horrid forms proceeded—worst perhaps in children killed, or maimed, or burned in the destruction of buildings or, as not infrequently happened, trampled in the movements of panicking crowds.

My mother received a letter reporting the experience of a nurse in a casualty station who had to communicate with the stretcher cases as they were brought in:

> It is ghastly having to go round as quickly as possible with the doctor trying to get out of their dying lips information as to identity, and some last message. In many cases their faces are blown away, as also their arms and legs. She says it is so pathetic to realise how anxious they are to give a last word before losing consciousness. The doctors and nurses often feel that it is about as much as they can stand, when another ambulance arrives and it begins all over again. Then a poor man comes up and says that his wife and five children were in a shelter, and that he has found two dead at a hospital. And we have to break the news to him that we have three children in the adjoining mortuary.

It is not surprising that during these days men and women should raise questions about the divine purposes, when so much physical anguish was meted out to so many and mental anguish to almost all. In response to much perplexity, Les preached a series of sermons on the will of God: he begins by quoting a woman as saying, "My boy was killed ten days ago in one of the raids on Berlin, but I am trying to bow to the inscrutable will of God." "But," said he, "was that the will of God? I should have said it was the will of the enemy, of Hitler, if you like, of the evil forces we are fighting. Are they the same thing?"

The sermons on this subject were later published as a small book, and they were very widely read. The book was still popular ten years after its publication when Les taped it in New York for the John Milton Talking Book Magazine for the blind. By the mid-seventies it had sold over 300,000 in the States alone. His argument about God's will requires that it be thought of in three parts: the intentional, the circumstantial, and the ultimate will. Their differences and their relationship can be illustrated by the crucifixion. The intentional will of God can be defeated and indeed *is* defeated by evil, but only for a time: it was not God's intention that men should kill Christ but that they should follow Him. The circumstantial will is that purpose which takes effect consequent upon the circumstances of human evil. Given these circumstances it *is* God's will that pain and death shall be inflicted inasmuch as the laws of the universe by which they are effected are an expression of His will: "the laws which govern the hammering in of nails held on the day of Crucifixion . . . If nails are hit with a hammer wielded by a strong arm they pierce the flesh even of the Son of God." The victim of circumstantial will might, however, in spite of pain or loss or frustration or grief, dominate the situation and wrest a spiritual good out of evil, as Christ did. The ultimate will of God was itself such that evil was to be rendered into good. And nothing could *finally* defeat that purpose, "With evil intent," says Les, "men crucified the Son of God, and within six weeks other men were preaching about the Cross as an instrument of salvation." So in those latter times of evil, ultimate purposes were being worked out, whether men had a glimpse of what they were or not.

Parts of this message were published in the press during these days, and they brought, as so often, a strong reaction. In particular, Les's treatment of Psalm 91 was to produce dissent. The words in themselves are superb: "A thousand shall fall at thy side and ten thousand at thy right hand, but it shall not come nigh thee." It would be good if it were true, as many believed, that prayer was the premium of an effective insurance against bullets and bombs. But it could be a painful exacerbation of grief for the bereaved to feel that they had failed in their prayers to uphold their beloveds. "The time has come," Les wrote, "when some of us must say quite definitely that that statement is not true. If you reply, 'But it is in the Bible', I nevertheless repeat that it is not true . . ." The editor of the paper called it "one of the most forthright and frank articles ever published in a Sunday newspaper"; but some readers objected strongly to what they considered a high-handed treatment of the text and the undermining of faith in the psalm, in which hundreds of thousands—soldiers, sailors, airmen and civilians—had found comfort and strength.

My father's postulates about the will of God may seem straightforward and unambiguous (they seem more so in summary). But he didn't attempt to evade the fact of the colossal mystery of pain: his long preoccupation with suffering had not by any means even blunted its appalling and tantalising impact upon him. There were days during the war when, as he records, for a sensitive man who suffered with every new story of tragedy that was told him, life in London was simply not worth living. And once, talking with a man around Christmas-time in 1940, when he mentioned a mutual friend who had died and the man said, "Lucky fellow," Les was startled, because for days that thought had been uppermost in his mind. There were moments like this for him. And yet that same winter saw the publication of a great ringing proclamation of his faith in God, *This Is the Victory*. In the teeth of the *blitzkrieg* he had gathered together his beliefs and his arguments and had marshalled them against the spiritual defeat of despair. "So from this dear old London," he writes, "now in an agony such as it has never known in all its long history, goes forth this message: that our true victory will depend on our faith in God and His purposes in the world. 'This is the victory that overcometh the world, even our faith.'"

There were some bright moments. In February of 1941 in his letter in the church magazine my father duly reported the arrival of spring flowers. And that same early spring brought also the great relief of the liquidation of the church debt. This was standing at more than a thousand pounds; it had existed since the outbreak of the First World War; and now, a peeress, begging to remain unknown, wrote a cheque for the outstanding figure, thanking my father for his sermons, saying, "I want my thanks to take a practical form": to buy off the City Temple debt seemed, she said, "more important at the moment even than investing in Government stock . . ."

Les exulted over the relief of this burden. But the euphoria was not to last very long: only six weeks after the thirty years' debt had been cleared, the old sanctuary was burned to the ground, leaving only the walls and the tower standing. Les was on the scene before the fire was out; there, seen from Holborn Viaduct, was old Parker in marble, smudged but proud; and there too, in the flesh, a girl from the church, standing alone, weeping. The outer walls stood and the pillars that had supported the gallery, but they enclosed only some heaps of calcined stones, strewn across with charred beams and a snarl of cables. My father's last sermon in the building had been titled, "On Making a Pact with Death".

But this was not death: the church, as Les had often insisted and now repeated, was not made with hands: it was indestructible. If violence and hate could destroy it they would have done so in the first century A.D. Now he exhorted his congregation through the church magazine:

> It is invisible and invincible, indestructible and indissoluble. Smash the form of its expression here and it breaks out there. Burn its local habitation to the ground one day and it has found another the next . . . So, resolute and determined, with quiet hearts and steady nerves, *and with no bitterness*, remembering the thousands of our fellow Christians who through all ages, have been driven from pillar to post for Christ's sake, THE CITY TEMPLE GOES FORWARD INTO ITS FOURTH CENTURY.

8

For some weeks the City Temple congregation gathered and worshipped in the Memorial Hall, part of the premises of the Congregational Union in the City. During the nights, war still raged overhead, and on Sundays it was sometimes difficult to approach the Hall. One bright spring morning Les drove down to the City which was still covered by a pall of smoke and lit by a filtered yellow twilight. He parked in Smithfield Market, and picked his way through streets half blocked with rubble and craters. As he walked he saw part of the Old Bailey, the Central Criminal Court, come tumbling down into the street. Fleet Street and Ludgate Hill were both impassable because of the heat of the burning buildings on both sides. At the same time another church member was finding her way through the rubble, her hair full of charred paper and her lungs full of smoke. She records how a policeman asked her where she was going. "I replied quite truthfully, 'To church'. He looked at me in a dazed kind of way and being too exhausted to argue, poor fellow, he let me through." Few people in the London area had slept better than fitfully the previous night, many not at all, and yet the Hall was packed with people. Les had to raise his voice in the sermon to be heard over the noise of fire and the hiss of firemen's hoses across the street. As he spoke a gas main exploded and the flash lit up the faces of his congregation with no divine illumination. He could see through the window as he spoke a burning building and, still undevoured, a sign "Hot Sausages Ready" which, it crossed his mind, might well have been the only understatement in the history of advertising!

Within two weeks of the destruction of the City Temple he received from the Reverend George Salter, vicar of the Church of the Holy Sepulchre, an offer of the use of this building for Sunday services. This Anglican church stood within a quarter of a mile of the Temple at the corner opposite St. Bartholomew's Hospital. Salter was the hospital chaplain and conducted his own services in the hospital chapel. The congregation of St. Sepulchre's had dwindled on account of the war: Les made it very clear that it was only due to the war that the church was available. "We are not 'taking over an empty Anglican Church'," he wrote, as suggested by "unintelligent comments in the press". But pretty soon he had got to know George Salter well enough (or so he assumed!) to tease him about the declined attendance. "You have to be careful to read the right service," he said, "or the congregation will come away married to each other!"

The offer of the church was of course eagerly accepted, and the Temple people worshipped there for five years, with occasional interruptions dictated by the need for repairs after bombs had fallen close by. Les naturally enough saw in the arrangement a gesture and a step in the direction of church unity. The Bishop of London and the Archdeacon had approved it, and the Dean of St. Paul's had agreed to lend chairs from the Cathedral to supplement the seating. He was delighted for one thing at the brotherliness of the Anglicans: all his career he had worked for church unity, and had urged it up and down the kingdom in sermons and lectures. Here, however minor, but not so minor at that, was palpable action which would do more, he felt, than many committees. He took the occasion to point out to his people his own role in the unification of denominations: he had been baptised a Presbyterian, trained as a Methodist minister, preached to Congregationalists in an Anglican building, and had, he said, arranged to be buried by a Baptist!

There was all that which pleased him about the arrangement. But there was also the history of the church that he liked, and there was his most un-Presbyterian delight in its beauty—beauty that the Church of England had deliberately fostered as an embodiment of holiness for 300 years, the same kind of beauty that had stuck in the throat of Cromwell and, no doubt, of his chaplain, Thomas Goodwin, who had founded the City Temple! The vicar showed Les round the church and spoke of the 800 years of its known history. Relatively late, it had been damaged in the Great Fire of London in 1666 and repaired "a great deal" by 1668, as recorded by Samuel Pepys in his diary. William Harvey, who discovered the circulation of the blood, had been baptised there, and so had Roger Williams, first Governor of Rhode Island. Here, according to tradition, Roger Ascham, tutor to Queen Elizabeth I, is entombed, and also the English adventurer, John Smith, the colonist, and, according to some, the Indian princess Pocahontas, who had saved Smith's life in Virginia, had married him and had later died down the river at Gravesend. On Christmas Day 1778 John Wesley had preached in the church. The vicar related these and other passages of history that the stones of the building had witnessed. He added, "In the evening on a weekday you may see an old man kneeling in the pew at the back there. He's a ghost: don't let him worry you. He drops in sometimes to pray." Les never saw him. But one morning Alice Head saw a clergyman up at the altar who, she thought, was assisting at communion—until Les walked right through him.

When the catastrophe hit his church Les had responded, as men do, with courage and adrenalin. In a way, when the thing that has been long dreaded at last actually comes to pass there is a momentary relief: it need no longer be worried about; it has happened. When the Temple was bombed Les could declare with a rallying cry the invulnerability of the spiritual church, and in doing so he could strike a chord in the hearts of many fellow Christians; he publicly recalled to them the Church's legacy of affliction, and, reminding men and women of the infant church in Rome that had suffered under Nero, he brought them into closer touch with their heritage. Materially the forties were dangerous times in which to be alive. For men of the cloth, however, they were heady times of unusual opportunity; now, with death always at hand, there was an appreciable growth in the interest in religion: Les used to quote Meredith: "Not till the fire is sinking in the grate/Look we for any kinship with the stars." Meeting the challenge, he did much travelling around Britain from the Gwennap Pit in Cornwall—the huge amphitheatre where Wesley had preached—to Edinburgh in the north, and he spoke of the new age that might be born out of the present struggles. Early in the war he began to speak of peace and the need to be spiritually prepared to "win" it, in the expression they used to use.

The buoyancy that appeared to keep him on top of this—preaching and praying and meeting the needs of the times—was maintained, however, only in despite of lingering poor health. In the winter of 1937–8 he had completely broken down, and he had scarcely recovered when the September Munich crisis, the mounting tensions of the months prior to war, and then the war itself were upon us, draining again the resources of strength that had only begun to return to him from their recent depletion. I don't think he ever got back absolutely the carefree health of his thirties. There were occasional days, sometimes a week even, when he felt good—perhaps after a day's walk in the country or some other absorbing excursion. But there were months together when there was no relief from the chronic headache that had dogged him, off and on, since 1937. It was not easy to carry a message of faith, hope, and good cheer from day to day and Sunday to Sunday. But he complained very little: he had inner resources, as they say, and he had a sense of humour and proportion which enabled him to think of his personal troubles as minor. We drove him to Broadcasting House one early morning and, though I don't clearly remember the sequence of events, we had parted from him when there was a delayed action bomb—a terrific roar, a cloud of dust, and, when this had settled, Les, peering across the street, looking for us, then grinning and shrugging his shoulders as if to say, "What the hell, it's all part of the day's work!" The title of the series he was broadcasting in was "Lift Up Your Hearts".

In October of 1941 he was invited to go to the States by the Federal Council of Churches and the World Alliance for International Friendship, for a cause which, as the titles of the organisations suggest, was dear to him. He accepted. But when he saw the itinerary he knew that he would not be able to muster sufficient strength to go through with it, and Lord Horder warned him once again that his reserves were low and told him not to undertake it. So he stayed at home and felt he had let them down.

It was not only the dull pain in itself, endless as it seemed, that afflicted him; he began to fear that it might be a symptom of some serious physical disease. All kinds of infirmity paraded before his imagination, which in turn was not slow to produce symptoms appropriate to this or that ailment. He went to various specialists who instructed him variously to drink more fluids, to drink fewer fluids, to relax with a cigarette, to stop smoking, to get new spectacles, to use a smaller pillow, to take a sedative, or to avoid woollen blankets. In the end he underwent a long psychological analysis of the kind he himself had formerly administered to others, now dredging up what was buried in the depths of his own memory and recording his own dreams. He gained, for sure, some self-knowledge through these long sessions; but the headaches persisted. The drawer of his bedside table was crammed with pill boxes, little tins of lozenges, and bottles of miscellaneous remedies. Once, opening it and surveying the miniature pharmacy, he remarked, "Not bad, is it, for a chap who believes in spiritual healing!"

He derived more therapy, I believe, from walking in the countryside. While the bombing went on in and around London, what summer holidays he could get he took fairly close to home. There was the Farmhouse School in Buckinghamshire where the family spent some weeks one summer. Bucks is an attractive county for its hills, its beech woods, and farmlands. There are narrow lanes between tall hedges, tangled with woodbine and honeysuckle, and there are many public footpaths, clearly signed. The Vale of Aylesbury where the farm lay was a good place for walks. It was good too for birds. Some years before the war Les had caught from Ernest Appleyard an interest in birds, which our whole family came to share. When we were in London, my mother would buy little cartons of mealy worms from the Army and Navy Stores, and with one or two of these delectable morsels she could entice the local robin to come in through the french windows on to the breakfast table. My father was very keen on watching birds and finding their nests. We used to mark them on their legs with small aluminium rings which bore the address of the British Museum, and it was some excitement to hear where they had been found and to learn their migratory movements. I think he liked best the birds of the moors and the wild country, like the curlew and the lapwing, not least for their romantic melancholy cries. But in the spring and summer in Buckinghamshire there were lanes where a whitethroat might come spiralling up out of the hedge, singing, or a greenfinch whistle through the heat; and sometimes the woods were loud with birdsong. Later when Norah Laycock made him a governor of the Farmhouse School, he was not in the least unwilling to travel out there for meetings. She was once overheard to remark, however, that his main interest in meetings was to get them over and have a good walk in the country and a meal with the farm products at the end of it. She wasn't far wrong at that.

When the bombing subsided Les took the family, or rather whatever bits of it were available, up to Moffat in the Scots Border Country. We were not, of course, allowed the use of petrol for our own car, but Les would order a taxi to take us out into the hills and then to come again in the evening to meet us at another point. It was Adamson the milkman who supplied this service—he who had also supplied the comment on Les, quoted earlier, that would serve as an epitaph: "He kens the tops." He had one of those old Austins, the kind you could climb into and stand up and walk around in, built before some fanatic decided that cars should be streamlined, and it would gently chug up the valley to Bodesbeck Law, or Hart Fell, or the Grey Mare's Tail.

Here in the mountains Les had his best days. His worries ebbed in the presence of these great permanent masses which he was never able to survey without associating them with lofty moral qualities in man and, as we have seen, with the Maker of the Universe Himself. The Grey Mare's Tail, a waterfall that feeds the Moffat Water where he had fished as a boy, was one of his favourite hikes. The path climbs up close to the burn and the waterfall, and then at the top it leads up to the edge of Loch Skene, where, according to Sir Walter Scott, "the eagles scream from shore to shore". To climb up and to come in sight of the loch, even if it were leaden grey and beaten by rain, would give my father tremendous joy. He would emerge from the canyon where the burn flows, swing his stick aloft, burst into "O lucky Jim", and start feeling for his sandwiches.

Then at the end of the afternoon we would go puttering back in Adamson's big square Austin. Les would head for the hotel bathroom, take a hot bath and then come down to dinner, neatly suited. He knew the formula for a good day: it would contain a long walk, followed by a hot bath and a good dinner; then cigarettes on the terrace, a game of bowls, perhaps, and early to bed with a good novel. There were days like that now and then. And as he became more analytic about his life he would record happy days by marking the date in his diary with a ring in green ink. A very happy day got a double ring round the date.

There were few enough such days during the winters from 1940 on. There were the grievous human losses of the war, and, remembering them, it seems facetious to record what was in truth the case—that for many people for much of the time the war years were dreary and boring. Food was dull and relatively scarce; restaurants served Woolton Pie, about which, learning that it was named for the Minister of Food, one was too reticent to enquire any further; cities were dark; shop windows were reduced to smallish apertures; one even missed the old, gaudy, vulgar neon. The services of trains and buses were curtailed and placards on the railway platforms asked, "Is your journey really necessary?" Petrol for recreation was completely cut off and posters admonished drivers to "Go Slow" to save tyres. Les was allowed a little petrol to get to the church and to visit the sick. And he knew a retailer who would sometimes give him a little more than the value of the coupons. "Are you in a four-for-three mood?" he would fondly enquire, smiling and exhaling that charm which even a tight-lipped civil servant with a narrow legalistic mind could hardly have resisted for long, and four for three would duly be delivered. But you still couldn't go anywhere you wanted. Les longed to be able to drive out of London and to get back to the Dial House, but the coast was closed and the house was commandeered and full of soldiers. Life had closed in on him like a winter evening, and on everybody else, for that matter. One ground-floor room of our London house had been shored up with timbers, and when nights were bad they would huddle together down there and listen to the thud of the guns. The war was no high drama: fuddled, half-awakened brains, bad tempers and boredom presided over these nightly vigils. At three in the morning when the firing had tailed off they went back to their cold beds.

Les's most prized possession those days was an interleaved Bible. My mother had had it bound for him in green leather, and for every page of print there was a blank page facing, on which in tiny letters in green ink he had written his own commentaries. As time went by it came to represent years of thinking, and its value became immeasurable. Every night during the war he would bring it down from the study at the top of the house and put it in the steel safe, proof against fire and destruction. It amused him to imagine the safe being burgled. "I'd like to see the look on a chap's face when he broke open the safe and found only a Bible for his pains."

Two developments in the grey war years brought Les a great deal of pleasure at the time and in following days. In June 1941, Winifred Haddon became his secretary, and Alice Head became a member of the Church Council. They both did much for his morale. Winifred Haddon was efficient and cool. She installed a green ribbon in her typewriter and learned quickly to deal with the enormous correspondence, sorting mail into its priorities, summarising some of the long letters and providing polite responses to the invitations Les received and could not accept. She typed sermons, taken down verbatim in the church Sunday by Sunday; she took dictation off the dictaphone, one of those old models with wax cylinders the size of beer bottles. Sometimes in the war she would take an hour off and go foraging in the town for the kind of boiled sweets Les was fond of. She was thoughtful and affectionate, but perhaps her greatest quality was her sense of humour. We credit people with this faculty when they gratify us by laughing at our jokes, not when they make their own; and we deny it to those who don't think we're funny when we mean to be. Winifred seemed to enjoy Les's humour genuinely, even when it was mordant; and she endured, possibly even with pleasure and certainly with good will, his endless teasing. Like anyone else who had anything to do with our family, she was shortly endowed with a nickname, Bunnus, which must somehow have derived from the rabbit, which she in no way resembled. You never know whether people *really* mind, and I sometimes think she was even more good-natured than we all assumed. She was a tall slim girl, her hair dark, carefully styled. She dressed well in the forties mode, and was always beautifully groomed. She brightened life for Les immensely. Years later he referred to her as his "all but perfect secretary for twenty years".

Sometimes she and one or two girl friends would come up to supper after the Sunday night service. We had few visitors to the house for meals, but Les found it relaxing to have company come in after he had preached. Also, on Sunday nights, Gertie was always out at Donald Soper's church, and Les would sometimes say, "If we bring some girls home to supper, they'll do the washing up." And they would. And then he would light a cigarette and slide far down in his chair while Dick put on records of Sibelius, Tchaikovsky, or César Franck and the atmosphere became purple and moody. Women's Lib was a quarter of a century away.

The Church Council was composed in those days of a dozen or so men and women. Some were elected by the church members and others appointed by the minister. Les appointed Alice M. Head, who at the time was the editor of *Homes and Gardens*. She had had a remarkable career as a journalist, much of which she recorded in her book, appropriately titled *It Could Never Have Happened*. At an early age she had been associated with the London literary world, having been secretary to Lord Alfred Douglas. But it was later, when she was working for the English editions of the magazines owned by William Randolph Hearst, that her career began to move among phantasmagoria, and particularly when, upon the death of her superior in the office, she became the editor of *Nash's* and *Good Housekeeping*, and Hearst's chief agent in England. One substantial part of her duty was to buy antiques and art works on behalf of her chief, who was a tireless collector. She would receive cables instructing her to buy certain lots as numbered at Sothebys' and Christie's. One day she had received a cable which read simply, "Buy St. Donat's Castle." She looked it up in a gazeteer, identified the agents, offered 120,000 dollars, and closed the deal, all within as short a period of time as if it had been a Sheraton chair or an ormulu clock. "I like to do as my employer wishes, so I bought it," she wrote quite simply. She had never seen the place. Nor had Hearst. The next fifteen years during which she worked for Hearst were not short of marvels. The extravagance of his way of life amazed her. She crossed the Atlantic and the continent literally scores of times to stay at San Simeon with him and with Marion Davies. She had purchased tapestries and miscellaneous furnishings for the ranch, as it was called, and two giraffes for the zoo there. Some of the tapestries had been coveted by Field-Marshal Goering, who, misjudging his military ability, told her in the summer of 1939 that he was coming over to England to get them.

Such had been her career. When Les came to London in 1936 she read of the crowds he was attracting and contacted him by a letter, commissioning an article for *Good Housekeeping*. On account of family ties she had been accustomed to spend weekends out of London; but now in May 1939, with the threat of war, she found herself at home on a Sunday for practically the first time in her life, and out of some curiosity she set off to the City Temple, without any particular feeling of keen anticipation.

> But [she records] the moment I heard Mr. Weatherhead's voice he kindled an instant response in my own consciousness. All that I had heard about him surged up into my mind and I thought to myself, "No wonder the people wait for hours to get in. This man has an

> extraordinarily attractive personality which gets across quite effortlessly and indeed unconsciously, through the medium of everything that he says." Seeing him in the pulpit for the first time, I knew at once that he was a man of genuine humility, deepest sincerity, and possessed of a true devotional spirit . . . His sympathies are so wide, his humanity so deep, his speech so direct and crystal clear, that no one can listen and remain unimpressed . . . I went again in the evening, standing in a long queue on Holborn Viaduct in order to get in. I hadn't stood in a queue for anything for so many years that I began to feel slightly irritated. Yet the anxiety to hear the minister again was so keen that I knew that I should stand in a queue Sunday after Sunday for an hour if necessary until the time came when I could secure a sitting at the City Temple.

Shortly she did secure "a sitting", though the church treasurer, instructing her that it would cost her ten shillings a quarter, enquired gravely whether she thought she could afford so much!

It is true and to her credit that Alice Head never looked like the highest-paid woman journalist in London nor yet the most brilliant, as she probably was. She knew a great deal about the kind of humility with which she credited my father. And it always amazed me, knowing of the glitter of her days in the Hearst empire, to observe the zest she had for simple things and the vigour with which she addressed herself to self-imposed duties associated with the church. She was fond of cats: there was Rufus, the church cat, who disappeared at the time of the bombing. Alice knew him well because he had taken part in some of the rituals that had been held in the church basement—he had been present at hymns and had once jumped on my father's shoulders when he was pronouncing the benediction; she grieved at his absence and was anxious to know what had become of him; and it was a great day for her when she learned he had joined the printers of the *Evening Standard* and had comfortably disposed himself in their premises. And then once in a small town in the north of England, when I was driving her and our family up to Scotland, she suddenly gripped my knee and exclaimed, "Mind the puss-boy." And there, about a mile up the road, a diminutive grey creature was standing on the kerb, looking as if it might be contemplating a dash to the other side. Alice had recently returned from California where she had stayed with Hearst and had been partying in Hollywood after their style; and we were in one of those stern limestone towns in Yorkshire and she was anxious about the puss-boy up the street, because, as she put it later, "They always run off in all directions." But that was later. Now, in 1941, my father was just getting to know her and appreciate her commitment to spiritual things and her great practical ability. It must have been the following Christmas that she sent him the fox-fur rug with the fox-tails hanging from it, explaining that Mr. Hearst had inadvertently sent her two of them.

To Winifred Haddon and Alice Head, Les owed much in the tough years of the early forties. There was also Albert Clare, who most conscientiously earned a place in the record. He had been associated with the church since the days of Parker and had himself looked after much of its administration. He was a short man: he had a face exactly like that of Sidney Greenstreet, almost as much suavity, and only a little less embonpoint. He was the editor of the church magazine and the author of the history that came out during the tercentenary celebrations. His prose, based on the style of Sir Sidney Robertson, was dignified; it was occasionally invaded by a massive witticism, the approach of which could be seen from afar. Les called him "Mr. Eclair" at first and then, when he thought he knew him well enough, "Pooh", after the amiable rotund bear in A. A. Milne. Though not offended, Mr. Clare wanted this otherwise, and then Les reverted, I believe, to "Bert".

Les had an awful nerve. Occasionally, as I have noted, people got mad with him—gardeners and housemaids had to be replaced—and really it is a bit surprising that it didn't happen more often. He was the youngest child in his family and retained much of his boyishness. W. H. Auden, the poet, once remarked, "I was the youngest child in my family, and in any company I always feel I'm the youngest", and I think this was true of Les to some extent. But then, with the mischievousness, there was this overwhelming charm which drew men, women, and children to him. My youngest child was four and she had never seen him when the taxi pulled up outside his house; and yet, as the car stopped, in one movement she opened the door and literally leapt upon him where he stood on the pavement, and lodged herself in his arms like a koala bear in a sweet-gum tree. Other people, without performing such a Nijinsky-like feat, were just as strongly attracted to him. He was a presence. Raynor Johnson in his book *The Light and the Gate* puts the matter quite simply: "Everyone felt that life was very good when he was around." So people co-operated when he smiled and asked if he could have four gallons for three coupons, and they forgave him when he used his own home-made nicknames.

After the bombing, Alice Head, Albert Clare, and Les began almost immediately to talk of rebuilding; and Les, ably supported though he was by these two and by other business people in the church, himself organised the plans. We have seen him as somewhat detached from mundane matters, not wanting to know about the minor details of the family alimentation, wanting to think about mountains rather than learn whether the laundry had fulfilled its commitments concerning the shirts. But he was not in the usual sense impractical and otherworldly: he was well organised and liked organisation: he could remember statistics; he kept lists; he knew his bank balance and paid bills as they arrived with cheques that were invariably honoured. On his desk he kept a sheet with five columns headed respectively "Write" "Phone" "See" "Do" "Get", with items duly checked. In his bedroom drawer he kept a card with a list of things he must pack when he went away to preach—an old worn card, written in the round childish hand of one of his earliest secretaries, listing such needs as clerical collars, aspirin, and window wedges for the rattling windows of the guest-rooms of old country manses, and later, in pencil, added in his own hand, "Spare teeth". He liked to plan things: he booked tables in restaurants; he never left theatre tickets to be picked up at the door, but had them in his wallet, ready; he checked times of trains, ordered taxis, and always got himself to the station on time.

He was a good organiser and there were, in fact, many business situations in the church in which he instinctively took the lead. Dick used to tease him about "taking over": he would describe situations in anecdotes which, relating some kind of mounting chaos, came to their climax with the remark, "Then Les took over!" Les would protest. But there was a grain of truth in it: sometimes his clear uncomplicated way of looking at things bore fruit in the world of affairs as it did in applying Christianity to life. Besides, he was lucky. Once at Brunswick, in Leeds, long after the Sunday evening service was over he found the stewards in the vestry still trying to figure out how much collection they had taken that evening. "Still here?" he asked. They explained that a shilling was missing somewhere, or two bob, or whatever it might be. "Well," said Les, beaming at them with a great air of demonstrating the obvious, "look in George's trouser cuff!" And then, riding the wave of this minor serendipity, when the money had been duly retrieved, "Good heavens, you fellows call yourselves business men! And it takes a parson, an idealist, a mere dreamer, to balance the books."

Well, Les liked getting involved in the business of the church, especially such radical and far-reaching business as building. And in 1941, after the bombing, he began already with Alice Head and Albert Clare to make plans for the rebuilding of the church, which was not to be finally completed until 1958, when after long delays it was opened in a ceremony attended by the Queen Mother and the Lord Mayor of London. Albert

Clare had prophesied in his rotund prose that the church "like the fabled phoenix would rise from its ashes in resurrected power", and so on. But Clare was dead by then, and what he saw of the new building to house the old church he had served so long was only a vision, as Israel was for Moses, from Pisgah heights.

Another personality who appeared, or rather re-appeared, to enliven the picture of the war years was Dean Lynn Harold Hough of Drew Seminary, in Madison, New Jersey. Hough had preached fairly regularly in the City Temple since his first appearance there in the uniform of an American army chaplain in 1918, just before the armistice. He had a captivating style in the pulpit and was successful as a preacher. He had been trained in philosophy and his thinking was largely transacted in the area in which religion and philosophy co-exist. "When religion touches philosophy," he once said, "it sets philosophical principles and sanctions on fire." He used to come over and stay in the Liberal Club, and he would have my brother and me to come and dine with him at the curiously early American hour of six. I see him now, a stocky figure with a huge briefcase, crossing the floor of the lounge and then settling in his chair with his legs folded under him. He would give us a box of expensive cigarettes, and he would tell us about Kierkegaard. In 1942 the U.S. State Department had been persuaded to grant travelling facilities to enable him to fly over and make his eleventh visit to the City Temple. He stayed in the Russell Hotel rather than the Liberal Club, since the club was near the river in what was considered a dangerous area, for the river was used by the bombers to guide them into London. My father thought highly of Hough, for his learning and his preaching ability, and he was not a little impressed by Hough's gesture of contributing his fees for eight Sundays to the Rebuilding Fund of the church.

But it was the church's spiritual role and not just the building that demanded most of Les's attention in those days. He broadcast quite often, and it was a thrill now for him to get letters from his audience on ships at sea, for example, or from the Libyan desert where two Tommies heard his voice as they perched on the hood of a Bedford truck, the radio set between them. Two officers heard a broadcast from Iraq—"ten miles," they said, "from a blade of grass"; and they heard also in the background what brought additional nostalgia—the roar of a London bus on Holborn, counterpointing the choir. Les repeatedly called for a spiritual awakening in England, particularly as he anticipated the rebuilding of Europe after the war. And the message gained intensity as the armies advanced.

In 1944, it was announced in the church magazine that on the third Sunday in October the evening service would be broadcast "if Hitler behaves himself"—this remark referring to the recent retreat of the City Temple congregation from St. Sepulchre's on account of flying bombs. Les exhorted the people not to sit back and simply let things happen. "A whole nation can be kept spiritually awake," he declared, "by the watchfulness of those who care, however humble and comparatively few they may be." He persuaded his people that "ordinary" people mattered. "Ordinary people won the war. Without 'ordinary' sailors, soldiers, and airmen, Wrens and Waafs and Ats, any opportunity of making a new world would have disappeared for our lifetime." He quoted J. H. Oldham: "What determines the course of history is the day-to-day behaviour of millions of ordinary people, acting in accordance with what seems to them to matter most."

One broadcast service early in 1945 was heard in Africa, and it brought the following message to Emil Mettler, a mutual friend of Les and of the sender:

> On the day of my 70th anniversary, in the evening, and at the stated hour by the B.B.C., we gathered in the room of one of the hospital patients, a European, who had brought with him a most excellent wireless set, so that he should not feel too lonely at the hospital . . . The heat was terrible. Sweat dripped from our faces and hands. Through the open window which is covered with a mosquito-net, we could see the palm trees lightly swaying in the wind, and heard the concert of crickets chirping in the grass . . . Suddenly, the loud-speaker stops scratching and we hear a tribute from the B.B.C., then a voice introducing the organ pieces, then the organ record itself, and a voice saying: "Dr. Schweitzer himself is listening at this very moment in Lambarene." I was deeply moved.

Sometimes responses to his broadcasts were less encouraging. Once he was holding a copy of one of his books in a bookshop and the assistant came up and said, "Did you hear him broadcast last week?" Les said he did and asked, "What did you think of it?" "Oh, *I* like him," said the girl. "But my friend thinks he's affected. His voice," she added. Les cruised off to the crime fiction section to pick up a Dorothy Sayers thriller.

He went to public meetings about peace, but I don't think he set much store by them. Then, along with other prominent figures, he sponsored the National Petition for a Constructive Peace which was to be signed by all or any British citizen over the age of sixteen and which emphasised the need for a peaceful order based on a "radical reconstruction—spiritual, economic, and political" and offered some detailed proposals. But I don't think Les believed that peace would be achieved by any organisation except the changing of men's hearts. That alone.

The end of the war came in Europe. On May 13th, Victory Sunday, Les preached to large crowds, morning and evening, which had gathered an hour in advance of each service. In the afternoon he processed with the clergy of the Church of England to the Victory Celebration Service in St. Paul's. On Victory Night he went down to join the crowds in Piccadilly Circus and in front of the Palace, and he felt the general elation. But there were memories that brought sad strains into the rejoicing: the dead, the bereaved, the crippled and blind, and those whom suffering and terror had driven mad rose up in his thoughts. And, even more sobering, there was the sheer responsibility of victory. "Victory," he wrote, "gives us an opportunity to begin to work for a new world, and defeat would have denied us that opportunity. But do let us above all remember that victory does not of itself change the hearts of men from selfishness to unselfishness and it carries no guarantee within itself of any improvement in the life of the nations."

V.J. Day came, bringing again thanksgiving and rejoicing, and along with it a good deal of feeling about the atomic bomb, ranging from misgiving to horror. Les was in London for the Thanksgiving Sunday. He had been reading in the *Encyclopaedia* the story of Richard Gatling, the American inventor who had perfected a machine which expedited the sowing of seed. Gatling had then turned in 1862 to the production of a gun that would fire 350 bullets a minute in rapid succession and had offered it to the U.S. Federal Government. The Government turned him down in horror at such a dreadful weapon. But now, according to the *Encyclopaedia*, "the machine gun is adopted by every civilized country". Les feared with some justification that shortly, in spite of our present horror, every "civilised" country would endeavour to employ the bomb. "There is no hope whatever," he declared, "in supposing that war can be ended by argument or by organisation or by fright. For one, I have no hope whatever of the future of humanity—let alone civilisation—apart from one tremendous fact for which this church has stood for three hundred years, that is the endless grace of God and the power of God to change the spirit of man."

It was towards the end of the war during the era of flying bombs and the V2 rockets which came over London with the roar of an express train that Les brought out *A Plain Man Looks at the Cross*. It is a series of meditations on the crucifixion in the characteristically clear style which makes the significance of this central Christian mystery available to laymen. It may have spoken to "plain" men, as intended. But my father, I think, regularly overestimated the intellectual staying-power of plain men. Visiting the British army of occupation in Germany just after the war, he was appalled to learn that most of the men spent their leisure time lying on their beds and gazing vacantly into space. And once, I remember, in some lonely spot in the Moffat hills, a shepherd pointed out his hut miles from anywhere, and when Les said, "I expect you get a lot of reading done in the long winter evenings," he was mildly taken aback at the reply, "I just set." The man then added in a spirit of compromise, "Och, the whiles I'll get a buich." The *Church Times*, never generous to Les, complained of the *Plain Man* that the author was unfamiliar with the Bible. But he knew it well enough; it was not lack of knowledge, it was his deductions that were unpalatable to orthodoxy.

During the war he had added appreciably to his steadily growing bibliography. Early in the war there had been *Thinking Aloud in War Time* and *This Is the Victory*. Later came *Personalities of the Passion, In Quest of the Kingdom, The Will of God, A Plain Man*, and *The Significance of Silence*. Some of these consisted largely of sermons he had preached, though they had all been rewritten and much of the direct address had been changed. If there was, as always, coolness from the orthodox about his books, many reviewers received them with enthusiasm. *The Will of God* was reviewed in the *Church Times* in the same article as C. S. Lewis's *Beyond Personality*, and readers were instructed that Les "was small beer after Mr. Lewis", a remark Les used to quote to amuse his friends. There were always objections raised against his liberal views, which I suppose were a necessary condition for the extraordinary vividness he was able to create in the familiar biblical scenes. When, for one example, he describes the crucifixion from the point of view of the centurion in command, he takes the action out of the stained-glass windows, out of the ornate illuminated texts, the images of smoothed marble in the Catholic churches, the old context of the legal transaction, and indeed out of all theology; it becomes not "the crucifixion" but "a crucifying" to be felt in the fibres of the hands and the feet where the nails went in.

9

When peace came Les once again found himself looking for accommodation for the church. St. Sepulchre's had been loaned specifically for the duration; now it was needed for peacetime purposes, and the vicar with all his habitual courtesy instructed Les accordingly. For some weeks, then, he used to plod across the city looking at a hall, or a church, or a disused assembly room. The rebuilding of the Temple itself was still a far-off divine event, and even in the minds of those most concerned it was a project felt to be of considerably less importance than the construction of homes. At length Les received and gratefully accepted an offer from the Quakers of the use of their premises, the Friends' Meeting House, a large building right opposite Euston Station. Consistent with Quaker practices, the hall to be used as a sanctuary was bare and had no organ; it was constructed for a kind of service in which the "beauty of holiness" had no significant part; compared with St. Sepulchre's, with its formal ecclesiastical features, it was severe. It was probably closer to the taste of Thomas Goodwin, founder of the City Temple and Cromwell's chaplain, but it was further from my father's. The offer was generous, however, and the City Temple people used the building for a year, until at Easter of 1947 they accepted an invitation to meet in Marylebone Presbyterian Church, near Marble Arch, where they were to remain until 1958, when at long last they were able to resume services in their own building on Holborn Viaduct in the City.

When the war was over Les began to travel abroad. In 1946 he went to Guernsey, to Ireland, to Sweden, Norway, and Switzerland. In Sweden, he addressed the Congregational Assembly and the Methodist Conference, where Bishop Arvidson, who was to translate the sermon sentence by sentence, introduced him by saying, "We are very conservative here in Sweden. I have read many of your books, and I find that you are quite heretical at the beginning, but you always end by being orthodox. Would you please begin near the end when you address our people?" In Norway, he spoke at a number of crowded meetings, and he met fellow ministers who had been victims of cruelty during the Nazi occupation. There were many grievous incidents related to him, impressing him with the pain that this people had endured and the courage they had shown. One minister, as he preached his sermon, had watched the

Gestapo come arrogantly forward down the church and arrest his wife before his eyes. He paused, as for any interruption; then went on with his sermon.

With some foresight, the City Temple secretary, himself an insurance broker, had taken out insurance on Les in case his plane should go down or his ship founder. And in fact he did have a brush with death. But, characteristic of fate, the event did not occur, as had been feared, in the icy clouds over the Alps or the pitiless waters of the North Sea but in the next street in our suburb, when, having nothing much to do, Les had made one of those little insignificant sorties on a dull Saturday afternoon to purchase some lemonade and buy the *Finchley Weekly Gazette*. My sister was driving the little Morris 8, and Les, with his head down, was fiddling with some packages under the front seat when the car suddenly cut across the road, and, in his words, began to climb a tree. He sustained a six-inch cut across the forehead and a concussion, and he was taken to the local hospital.

While in hospital, he was visited by Winifred Barton, the minister's assistant at the City Temple, whose job it was to visit the sick in the congregation. She was a diligent and dutiful person, conscientious and staid. Les had been dosed with some drug that contained salts of opium and it had induced in him a state of complete euphoria. He wore a lovely smile as she came up to his bed. "I've found the secret of happiness," he said, quite simply. Now Miss Barton had listened to hundreds of sermons of my father's in which happiness had been presented exclusively as a spiritual condition, a product of a relationship with God; so she thought some spiritual secret had been vouchsafed to him and enquired most tenderly, "Do you feel that you could tell me about it?" "Oh, yes," he said. "Opium."

He was well cared for, and the wound healed; but the after effects were prolonged into the autumn and winter of that year. And a bitter winter it was—the worst in living memory. As always, with the cold there were strikes and power cuts and a shaken government. Emmanuel Shinwell was the Minister of Fuel and Power, and, inciting us to conserve energy, he let it be know with some frankness how infrequently he himself bathed. Winston Churchill had the most corrosive way of referring to Shinwell: "The Minister of Fuel," he would growl, and then, as if this unspeakable designation were, in itself, an insufficient ignominy, he would pause, clear his throat, and add in another growl, "and Power." Outside the House, people used to say, "Where did Winston get all that snow?" My mother, brooding over the kettle in the kitchen, filling hot-water-bottles for the beds in the icy rooms upstairs, would mutter,

"I'll give him Shinwell"—an ambiguous donation but calculated, one had to assume, to be injurious. It was an iron season. Les had always felt the cold and hated it: he believed that his blood had been thinned during his time in India and Mesopotamia. He hated the east wind, or for that matter any kind of draught at all. He would sneak behind the drawn curtains in the living-room and surreptitiously close the window that my mother, a fresh-air fiend, had surreptitiously opened. In the winter he used to wear a balaclava helmet in bed. In the bad winter of 1946–7 after the car accident, his vitality ebbed and he felt generally run down.

In the spring of 1947, however, there was some good news: the City Temple was invited to use the premises of Marylebone Presbyterian Church. This was the seventh building they had occupied since the fire, and they were to keep it for the rest of their exile. Shortly after they moved there, a Presbyterian minister, Herbert T. Lewis, who answered to the name of John, joined my father as assistant minister. He had three pretty daughters, and he was a sportsman with a dry wit. Les and he got on fine. They played golf together and Les was a good loser. Once they were disrobing after service on a Sunday, and John was shaking his head solemnly. "What's up?" Les asked.

"You said a terrible thing in your sermon," said John. Les was not entirely unused to such protests; sometimes in letters, sometimes in head-shakings as now, and sometimes in my mother's comment, "Oh, Les, you were *awful*!", he would be told he had said something offensive. "What've I said, John?" he asked. "Is it the Catholics, or the Plymouth Brethren, or the Christian Scientists?"

"All of them," said John. "You've just outraged hundreds of good honest-minded men and women." He went on to explain that in his sermon illustration Les had had two teams playing each other—Leeds United and Sheffield Wednesday, they might have been.

"Well?"

"Man, *they're not even in the same division*! They *couldn't* play each other!"

Lewis died of a coronary as he was changing out of his golf togs to take a funeral. And when he had been cremated, the ashes were poured out by the curate and, being caught by the wind, blew into the faces of the widow and Les. They looked at each other and laughed: they knew that John would have enjoyed that. But Les missed him. "No one," he wrote, "could have been so easy to work with or more lovable."

But that was later. Now, when they had all settled down in the Marylebone church, Les tried an experiment which he had had in mind for many years. He had felt that there should be modifications in the church service. He didn't believe that hymn-singing and preaching were anachronistic—though he certainly thought that *some* hymns and *some* sermons were worthless, and he believed that the lofty nineteenth-century rhetoric with which Joseph Parker had thundered at God, man, and the Sultan of Turkey had had its day. He did not by any means believe that prayer, private or corporate, had become outmoded. But like any sensitive being he recognised the currents of change and tried to respond to them. He was, as always, impelled partly by his desire to make Christianity meaningful not just to the faithful, the regular pew-renters, but also to the casual stranger who had dropped into the church out of uncharted motives rather than routine. So in October 1947 he devised a new kind of service. It was to be in four parts, each separated by an interval during which people might enter or leave. The first part was hymns and prayers: the second, the sermon; during the third part Les came down from the pulpit—in fact and in symbol—and answered questions which members of the congregations had written down and sent over by the stewards. The question period was the most interesting and important departure. But when the press reported on the plan it was the intervals rather than the content of any part of the service that made copy and, indeed, a sensation. The *Daily Mail* ran a headline, "Pop out of church and get a drink", and a number of other papers described the experiment as a service with tea and cigarette breaks. Les wasn't unduly exercised by the press, but he was grieved to receive a clipping from South Africa accusing him of offering drinks in church and, following this, a sad letter from an eighty-four-year-old South African lady pleading with him not to degrade religion in this way. The church was full to overflowing when this service was held, and he was pleased with the experiment and repeated it. Or, on occasion, he would simply devote sermon-time to the answering of questions. Questions were written, not put vocally, but he promised the congregation that he would take them unordered and unedited and would at least attempt to answer each as he pulled it off the pile. To some questions that could not be adequately dealt with on demand, he gave provisional answers; then later he would devote a whole sermon to the topic. A number of such questions he answered at length in his book *When the Lamp Flickers*, which appeared at the end of 1948. But he was jolly good at responding on the spot. He was quick-witted, and he was not fazed by tough theological questions, or questions about politics, sex, or simple social matters. They asked him about angelology rarely but the divinity of Christ often; was masturbation a sin, they wanted to be told; should Christians dance, or vote Labour, or bet on horses, or go to the flicks on Sunday? He said something to all of these enquiries. He wasn't fazed by rudeness, either; and once, picking up a question, he smiled broadly at the congregation and said lightly, "Well, well! I've received many anonymous letters in my life, which have a message but no signature, but never before a signature with no message. This simply reads, 'Fool'." He was a great admirer of repartee, and his hero in this kind of wit was Donald Soper, who used to address an open-air crowd each Sunday in Hyde Park and was magnificent in the retort-courteous. Les loved retailing these. "There was a chap with a bald head who kept waving his hat and interrupting 'til Donald said, 'You better put your hat on; there's woodpeckers about!'"

In April of 1949 Les sailed on the *Queen Mary* in order to deliver the Lyman Beecher lectures at the Yale Divinity School. He also preached and lectured in New York at the Fifth Avenue Presbyterian Church, at the Union Theological Seminary, and at St. Bartholemew's Cathedral. It was a strenuous tour. In all he spoke fourteen times in the sixteen days he was in the country, and then there were question sessions after the Yale lectures and, following these, long informal discussions with students lasting into the small hours, and many interviews with individuals. There were formal occasions in New York, where he met for the first time men and women whose work he had admired from afar for many years—the Niebuhrs, Helen Flanders Dunbar, and Harry Emerson Fosdick, whose early book, *The Manhood of the Master*, he had carried around with him in Mesopotamia in the First World War.

There used to be a generally accepted idea that public figures visiting North America from England would be so heavily programmed that they would suffer lasting injury to their health. Les's farewell to colleagues visiting these shores was pretty regularly, "Don't let them kill you!" He was himself to take heed of this kind of admonition some five years later when, arriving in New York for a preaching tour, he was prevailed upon by John D. Rockefeller to cut out large parts of his schedule; and even then he was severely strained by the part that he kept. In 1949, however, he rose tremendously to the demands and, writing from the *Queen Mary* on his return trip, he says, "I enjoyed every minute of every day and never felt more fit in my life." He was no doubt buoyed up by the very nature of the major assignment—the lectures and the spirited intellectual collisions with students, with whom he was thoroughly delighted. But also he was met all round by so many genial people who made him feel wanted. During the war an American minister had been reported as saying, "After the war we'll wait and see what Leslie Weatherhead has to say", and although I don't suppose this was the general feeling among churchmen along the eastern seaboard of the country, there was no question about the general reception Les received in 1949. The convocation at Yale was the best attended in twenty-five years and commanded more interest than any other that the Dean could remember. The Professor of Preaching, Halford E. Luccock, wrote with suitable enthusiasm of the enthusiasm Les had generated in the audience. "All felt the shared miracle of spiritual communication, the eloquence of Christian experience, the man in Christ sharing the richness of his faith and love, and the insights of his mind, in words simple and deep and moving. The response which the audiences made was more than gratitude for intellectual stimulation and vision. It was the persuasive conviction, 'Surely God is in this place.'" The ministers Les met in New York were likewise delighted with him: the audiences were unusually large and tremendously appreciative. He must have felt like a prophet out of his own country.

The mental boost he derived from this adulation was seconded, Les thought, rather mundanely, by food. For three weeks he had been liberated from the grey disciplines of British rationing—the Woolton Pie, the little piece of weekly beef the size of the heel of a lady's dancing-shoe, the powdered eggs, the powdered milk, and all the glum *ersatzes* to which we had been reduced. Now he was amazed at the food proffered—the enormous steaks, the roasts, the fresh vegetables and salads, the cream desserts. He was amazed also at the costs: his room at the University Club, Fifth Avenue, where he was a guest, cost, he learned, $11.50 a day; a steak cost $3.50—though now these figures merely make us wish we had eaten more in 1949!

What made the trip so pleasant for him, however, was largely that he was accompanied by his eldest son. Dick by this time was a qualified psychiatrist, thus sharing a great many of his father's interests. They always got on well, and one can well imagine how the tension of occasions could be relaxed by the presence of a close friend who, besides sympathy, had incidentally a medical degree and a psychological diploma and among all his professional credentials an unmatched ability, in his off hours, of teasing Les to the top of his bent and playing the complete buffoon.

Les came back from the States, then, full of beans—or rather, beef—and ready to go. Life was good. The church was flourishing—the church as a whole. Most of his energy as minister had always gone perforce into the preaching, but he was regularly concerned with the church's corporate life: he repeatedly called for attendance as an act of corporate worship irrespective of whoever was preaching, and he was particularly concerned with the life of the church organisations. Church members were encouraged to join these: there were prayer circles, as already mentioned, a choir, an orchestra, the leagues of service, and working guilds. What engaged my father's interest most, I think, was the Friday Fellowship which he himself had started. Before the war it had had a membership of three hundred and was the strongest element, Les felt, in the whole life of the church. It met on Friday evenings to discuss a previously arranged question—doctrinal, perhaps, or social, or other. For example, "When you pray, 'Be with me, Lord today', what do you really expect to happen? Does it happen?" People got to know each other and their minister in these evenings. Anything he had said in the pulpit could be brought up and questioned.

Another organisation he had started was the Samaritan League, a group of men who helped the poor, securing for them homes, meals, clothes, medical aid, radios, legal advice and pensions, visiting the sick, people down on their luck, dead beats, and prisoners. He had also started the Psychological Clinic, in which at one time eleven doctors gave their services. Payment of fees was invited, but no one was ever turned away. Over the years there were a number of sensational occasions when those who had been carried into the clinic walked out of it on their own and those afflicted by neuroses claimed they had been preserved from suicide. But there were many more unsensational occasions, when, as Les put it, the Master's work on earth was being performed here in the city streets. He also started the Women's League of Service, the Prayer Circles, the Sunday Night Social Hour, and an Adoption Scheme by which poor families were looked after by middle-class ones.

It was some credit to the general spiritual health of the church that most of these organisations persisted in their activities through the war and through the bombing, meeting perhaps in the houses of members or, most often, in the schoolroom of St. Bride's Church in the City.

The Literary Society had had to suspend its meetings for most of the duration. It had been a flourishing big weekly meeting with distinguished speakers from all areas of human interest. Les himself had lectured to it on a number of occasions for some ten years before he became the minister of the church. In 1948 it reopened and began again to attract large audiences which included the press. Les often used to chair these evenings and got a good deal of pleasure from meeting the more interesting figures on the programme. One such was the Radio Doctor, Charles Hill, a popular personality, who closed a lecture around Easter-time in 1950 with the declaration that the fear that was prevalent in the world was a significant source of general ill-health. The source of that fear was then Russia. And thus in thanking the lecturer, Les suggested that, as a gesture of the nation's goodwill, Princess Elizabeth and the Duke of Edinburgh might visit Russia so that their manifest charm might win Russian officials to believe in the British wish to be innocuous and friendly.

These remarks were made at a time when Joseph McCarthy in the States was raising his pernicious hullabaloo about the dangers of communist infiltration. But no one, even in those days of outspoken hostility to communism—when Christians across the whole spectrum of the church were declaiming against the antichrist of Russia, when expatriates enlarged on how and why they had chosen freedom, and when, in *The God that Failed*, half a dozen ex-communist intellectuals from the West explained in lengthy apologetics why they had deserted from the ranks—no one even in those days had spoken out against communism more firmly than Les had; he denounced it as absolutely incompatible with Christianity, and that was that.

It was a bit of a shock, therefore, when Lord Vansittart, in an hour and a half's oration in the House of Lords, was reported as having labelled Les a communist. In fact, Vansittart had merely denounced the scheme of using the royal couple as ambassadors and sending goodwill to Stalin, who had killed some thirty million people. In the same speech, however, he had named other churchmen as communists, and the press and popular opinion liberally distributed the honour to cover Les also. Within an hour of the end of the speech the B.B.C. had begun to receive phone calls demanding that Les be taken off the radio programme that had been scheduled for the coming Holy Week. It was very trying: a spate of letters and phone calls, some hostile, and some bewildered, came to the house, enquiring what it all meant. By the time the New York newspapers had got on to the telephone, Les was getting a bit irritated and he told them peremptorily to mind their own business; he would have hung up if they had not reminded him that, if he ever wanted to visit the States again, an immediate clarification would be useful. In view of his planned journey to Australia, where the communist paranoia was pitched pretty high and where in fact the press did report him as being on Vansittart's black list, the storm in the local teacup began to have wider implications and became a serious inconvenience.

In the end it all blew over: in the cold light of October, Les had a letter from Vansittart regretting the communist innuendoes in the press, disclaiming any such intention on his part, and affirming his high regard. Later they had lunch together and the damage was more or less made good.

The great respect that Les held for academic degrees has been noticed earlier in these pages. Before the war, of course, they were a good deal less common than they are now, and he never failed to include the letters of a degree after any name he might be citing. In the early stage of his career in the twenties he himself had earned a Manchester M.A. in English literature under H. B. Charlton, working for it while running his first English church, writing late at night, in the Cecil Street house, sharing the golden images of the romantic poets while coal-gas fumes leaked up from the cellar and the rain beat down in the dark street outside. The M.A. was the only degree he possessed until the fifties; he suspected and in one case knew for sure that honorary doctorates had been withheld from him on account of his book on sex. But he wanted a doctor's degree from a British university before he should be given an honorary American one, for it was then a fairly familiar practice for an English minister preaching in the States to be engaged to give a commencement address at the local college and then duly receive its Ph.D., rather as heads of state visiting China in recent years collect a couple of pandas apiece. He therefore began work on a London Ph.D. in the field of psychology under Eric S. Waterhouse, Professor of Philosophy of Religion and Principal of Richmond College. They had been friends for many years, and it was a curious round in the game that their relationship should become that of tutor and student—the tutor, the older man; the student by far the more distinguished, with twenty books already to his name. The tutor assigned essays to be written, and Les duly wrote them and brought them over to Richmond to discuss them; but the discussions frequently turned with immoderate celerity to railway trains—a subject even nearer to the tutor's heart than psychology. Waterhouse was a railway enthusiast and would recount such anecdotes of the lore as collectors hoard and relish. On one occasion he could report that he had actually got to drive the 10.34 from Paddington: this was his finest hour and his best anecdote and that day they read no more of the book!

Les was working on this London degree and had passed the qualifying exams some years previously when in the summer of 1948 Edinburgh University awarded him an honorary D.D. It was a splendid occasion of scarlet robes and purple hoods; Les gave an address during the celebrations, observing that the only person known to have declined a D.D.

was a certain Revd. Fiddle. When the affair was over, accompanied by his second son he took a boat from Leith to the Shetland Islands and spent a week watching the sea birds—the shearwaters, and the divers, and the gulls, and the great skua, that extraordinary misfit of creation that lives by terrifying other sea birds on the wing, inducing them to throw up their food, and catching and devouring the vomit in mid-air, of which indelicate practice Les's theodicy, though copiously illustrated from bird lore, never made much use.

The works on the thesis for the London Ph.D. progressed, and the degree was granted in the autumn of 1950, the occasion coinciding with the opening of his fifteenth year at the City Temple and the thirty-fifth anniversary of the completion of his first college course and the assumption of his first pastoral care. In the following summer the thesis was published, *Psychology, Religion, and Healing*, his first big book, dedicated to Arthur Dixon Weatherhead, with his string of medical degrees. It is a comprehensive survey of all the non-physical methods of healing. It begins with the healing miracles of Christ, and healing in the early Church. Then it deals with early methods of psychological healing: mesmerism, hypnotism, and suggestion. Then come modern methods of religious healing, including the laying on of hands, the healing phenomena at Lourdes, Christian Science, healing missions, and psychic methods of healing. Next come summaries of modern psychological methods as used by Freud, Jung, Adler, and McDougall, with a meagre page or two spared for Watson. Then in the next two sections come the main questions: Do psychological methods need religion? and Do religious methods need psychology? The questions were vital for Les, and the answers had to be Yes. Finally, as a corollary of the affirmation, comes the section on the co-operation of pastor and doctor.

Some of the ideas in this book we have already seen in other parts of Les's work. But there were new things too. Among the modern methods of healing, he finds that after much consideration he believes that certain people have a psychic healing power which can be brought to play by the laying on of hands. A minister he knew possessed this power.

> He was visiting the home of some of his people who were almost distracted by grief because their only child was supposedly dying. The minister went up with the parents into the child's room, and they prayed together. Then the minister, with deep feelings of pity and love, put his hands on the child's head without any thought of healing. Nothing was farther from this minister's thoughts than the laying on of hands. To his amazement, as soon as he laid them on the child's forehead,

his hands trembled and shook as if with a clonus. Medically and surgically, all hope of recovery had been given up. The child stirred, opened his eyes, and then, with a sigh, went to sleep. He slept naturally and peacefully for some hours, and then, with the speed at which little children so frequently recover, he sat up, wanted food and toys and appeared to have recovered. In a few days he was well.

Les was also firmly convinced that real cures of organic diseases had occurred at Lourdes, but of the thousands of patients who make the pilgrimage and bathe in the pool only a tiny fraction had been healed. He had himself visited Lourdes a few years earlier, accompanied by his elder son, travelling there as a pilgrim, on the pilgrim train along with hundreds of patients. He had misgivings about the pool, the slow exchange of the water and the incontinence, sores, and ulcers of those who were immersed in it. But as dusk fell, and thousands of pilgrims carried their candles into the great square and repeated the creed together under the stars, Les was moved. The Pyrenees were in the background.

In promoting spiritual healing Les had repeatedly emphasised that physical methods were indispensable: they were to be complemented by spiritual methods, not replaced. He was thus opposed on principle to Christian Science, and, in addition, some of the details of the career of Mary Baker Eddy, its founder, he considered ridiculous and so presented them. He aroused some animosity among certain of the Christian Scientists who claimed that he had used the wrong sources, and after their protest he corrected one or two errors of fact. Other people opposed his notions of demon possession, and yet others his interpretation of the healing miracles, charging him with an "astonishing ignorance of the biblical doctrine". The general reception of the book, however, was overwhelmingly favourable. It was recommended by both clergymen and doctors for students and practitioners of both medicine and the ministry; it was "the finest review of the whole subject in existence"; the treatment was marked by "a sanity which does not exclude courageous thinking". In particular the summaries of Freud, Jung, and Adler were praised, and the authority the book gained from both the scholarship on which it was based and the author's own thirty years of practical experience with psychology, religion, and healing. Even from the august Latinist heights of the *Times Literary Supplement* a crust of praise was thrown: "A positive contribution to the harmonisation of the various approaches to the relief of human suffering . . . [it] points the way to a more comprehensive attitude which may release the great energies of the non-physical world for the welfare of mankind." Its great merit, once again, and what won it such wide acclaim was, I believe, the clarity of the prose and, as this indicated, of the thinking behind it; Les had dealt with a great deal of complex material in which many men and women were vitally interested and many more prepared to be; and he had made it readable.

In the late twenties and early thirties Les used to go up to London from Leeds to preach and lecture. In 1928 he met Raynor C. Johnson, a lecturer in physics at the University of London, and subsequently he used to stay with the Johnsons when he made his regular London visits. They had much in common; for one thing, their humour ran on the same lines of mutual teasing and the ironic glance at solemnity. Raynor Johnson recalled a function at the City Temple where a number of distinguished churchmen had gathered. "Leslie was representing the Nonconformist churches and had made a hurried trip from Leeds to London to do this. It was a full-dress affair and no outwardly impressive feature had been overlooked. We had no opportunity for a chat but as we shook hands and exchanged brief greetings he whispered the message, 'Tell Mary when you get home that the sight of the day has been a bishop in his braces.'" And one summer when walking together in the country in Sussex they came upon a kind of clinic—some curious kind of massage and orange-juice operation—and by chance they fell into conversation with the director, of whom Raynor enquired whether he thought highly of Leslie Weatherhead's books and, on finding that he did not, he proceeded to egg the man on with encouraging nods until the books were thoroughly torn apart and Les was practically choking with suppressed laughter. It was an act right out of *Three Men in a Boat*!

Among their common interests, religion was, not surprisingly, paramount, and what appealed to Raynor was Les's controlling idea of it not as a system of creeds but as a relationship. In the end he was to pursue a career in mysticism and visions that left Les behind, but for thirty-five years they were friends, even though their communication after the first six years of their acquaintance became epistolary when, in 1934, Raynor Johnson moved to the University of Melbourne to become the Master of Queen's College. Then in 1951, having received many invitations from Australia and New Zealand over the years, Les obtained release from the City Temple officers for six months in order that he might preach for an extended time in Melbourne, at Wesley Church and the Congregational Independent Church, and then proceed home through New Zealand, Fiji, and North America.

In his favourite study chair.

With Dr. Coggan, the Archbishop of York, in 1968.

Opening the Leslie Weatherhead Room, 1971.

He sailed from England at the end of January 1951 with my sister Margaret, then aged twenty-three. They stopped in Colombo, Ceylon, where Les addressed an overflow crowd on spiritual healing. It was strange to be back in the East after thirty years and to dine at the manse and find among the guests a couple who had been at his wedding. After dinner, they sat outside, unprotected by netting; Les enquired about the mosquitoes and learned that they had all been disposed of by D.D.T. The ship put in for a few hours at Fremantle and he had time to go to Perth to conduct a midday service. Then at Adelaide he spoke in the town hall. And then finally they sailed into Melbourne, where he preached each Sunday in the Methodist and the Congregational churches alternately and gave lectures in mid-week. All the church services and the lectures attracted crowds; and once, arriving to give a lecture, he had to push through a crowd to get near the door which he then prised open by main force, only to be told, "It's no good shoving. You can't get a mouse in here." Since on earlier occasions many had failed to gain admittance, his farewell lecture was held in the Exhibition Building; it was attended by six thousand people. According to Raynor Johnson, no one from abroad except visiting royalty and Billy Graham had ever attracted such a following.

During his stay Les was criticised by some for his theology and by others for his lack of it; he was blamed for preaching psychology and for not believing in the story of Jonah. "It would be hard," he had said, "to find a whale suitable to accommodate Jonah for the weekend." He did some criticising on his own part: as we have seen, he was habitually most willing to bring Christian principles to play in the mundane and quotidian affairs of men, and it is not surprising to read in the papers of the day how he condemned the failure of the work ethic in Australia, the fact that men no longer felt proud of doing an honest day's labour. He also condemned drunkenness, saying he had seen more of it in three months in Australia than in three years in England, but adding facetiously, "Of course it is hard to get drunk on English beer"—a matter on which he was less of an authority than some of the papers, who took the flippancy seriously, seemed to suppose. But even given his conviction that Christian principles should be brought to bear in practical life, it is a little perplexing to read headlines in the Melbourne *Sun* reporting him as complaining about the trams and recommending trolley-buses; and one wonders which exegesis of which obscure text in Holy Writ has interpreted the latter as preferable.

On the whole the Australians thought very well of him. The *Age* deposed as follows:

> They laugh heartily at his witty sallies, they hang on every word as he plumbs the very depths of their souls. Some, indeed, lower their heads when he suddenly hits the mark, as if afraid to admit that he has discovered the secret they have been trying for years to hide . . . Young women, city typists, factory workers, shop assistants, many from rich homes as well as humble, are the Doctor's most profound listeners. He seems to have the answer to so many of their troubles . . . Dr. Weatherhead's message is the old, old story, but with this difference—it has a vigorous punch that startles men and women out of the years of religious apathy into which so many have lapsed.

The minister of Wesley Church, Irving Benson, wrote, "The supreme glory of his preaching is not that Dr. Weatherhead is a pulpit giant, but that men and women go home saying, 'Isn't Christ wonderful!'" Les cherished this kind of praise: this was what he wanted to do.

As the weeks went by in Australia he found that his time became increasingly engaged; on top of his scheduled preaching and speaking there were sessions with students and interviews with psychological patients, some of whom came a long way to talk with him. All the same, there were times here and there for excursions into the Victorian countryside: he saw exotic birds that intrigued him, and the koala bear, sitting in a tree looking like a little shrivelled woman, snoozing after lunch. Then he flew to Tasmania by night and from the plane saw big stars unknown in the other hemisphere, and Orion upside down. He saw seals cuffing each other like schoolboys and felt sure that the animals and the birds of the planet loved life and suffered only minimal pain.

It was not the prospect, however pleasing, of viewing the flora and fauna of these new territories that had drawn him here from the other side; in addition to the opportunities for preaching and lecturing and the persuasive invitations, there had been the hopes of having conversations with Raynor Johnson. These hopes were implemented: Les and Margaret stayed in fact in the Master's Lodge with the Johnson family as long as they were in Melbourne, though Les and Raynor dined in Hall each evening. Each had his own work to do, of course, but they found occasions when, as an earlier Dr. Johnson used to say, they could fold up their legs and have it out. Their discussions on religion were bent towards a practical end: they were planning to write a book together which was to commend Christianity to people who were dissatisfied by its conventional presentation in the churches. They got as far as a draft of the table of contents. In the end, however, they were to go their own ways: when

Les left Australia they recognised the difficulty of collaboration from a distance, and in any case their views seemed to be diverging. Les returned and became immersed again in his regular church duties; Raynor brought out *A Religious Outlook for Modern Man*. Then in 1964, he published *The Light and the Gate*, from which quotation has already been made, which speaks of Les in terms no one could wish to gild but concluded with judgment that simplifies and distorts a good deal of what was complex in him, that "he is too outgoing to find the mystic's way an easy one at this stage of his pilgrimage, for it involves considerable withdrawal from others and the subjugation of the restless mind to stillness." The next year, 1965, when in the freedom of retirement Les brought out *The Christian Agnostic*, the culmination of all his thinking on the subject that he and Raynor had discussed, Raynor wrote with little enthusiasm: "I feel very deeply," he said, "that all 'communication' by words (which are mental coinage) is strictly limited, so limited that I feel that where all the most important spiritual things are concerned they can convey nothing." The profound questioning of language and doubt as to its efficacy has been, of course, a characteristic attitude of modernism; but Raynor seemed now in these matters to have moved away beyond the last human habitations, where the suburbs give way to the unknown.

Les and Margaret returned across the Pacific calling at Fiji. As the plane landed here it was surrounded by chanting and singing natives, who had come out to the airfield to bestow on Les the Order of the Whale's Tooth, a ceremony traditionally performed in honour of visiting chiefs. The emblem itself was as big as a man's fist, threaded with a rough hemp rope to be worn around the neck. Les was very proud of this honour, since only three other Englishmen, one of them George V, had ever received it. It aroused curiosity in the press at home, where it was reported under the headline "Big Chief Returns" in the *Yorkshire Evening Post* and "Dr. Weatherhead's Tooth" in the *Dundee Courier and Advertiser*. But the Fijians had an even more moving tribute to pay him. One evening in the dusk they gathered, a sea of dark faces, under the balcony of the house where he was dining and began to sing. Someone had dug up a tune that Les had composed some half a century earlier, and now without warning he was suddenly aware of the singing from scores of voices welling up from the garden, with words to his own music.

He and Margaret flew across the rest of the Pacific, enjoying on the way two Wednesdays, and landed in Vancouver, British Columbia, where Les preached and lectured. He then went south to Tacoma,

Washington, to stay with his younger son, then a professor at the College of Puget Sound. Returning again to Vancouver he took the Canadian Pacific Railway across the Rockies—that most fabulous railway journey—to Winnipeg. Here Dick met him and motored him to Toronto, where he preached, and then on to New York where he boarded the *Queen Elizabeth* for home.

10

Often Les used to say, "My sons are my best friends," and sometimes he would add, "Not many men can say that." It was perhaps true—as true as any such generality could be: we both of us got on really well with him. He had other friends, friends of long-standing—Douglas Britten, Frank Dorey, Ernest Appleyard, Donald Soper, Billy Northridge, William Sangster—the kind of people one has known so long that one can be absolutely oneself with. But most of these lived outside London. Soper lived close. But when he saw Donald it was usually only a fleeting glimpse across a speaker's platform; in the twenty-five years they were both in London, they didn't play golf together more than half a dozen times, and I don't think Donald and his wife Marie ever ate at our house; but sometimes on his way home through Hampstead Garden Suburb, if Marie was in the garden, Les would stop the car and pass the time of day with her—he liked her very much.

He treated us boys like adults for the most part. Very early on, before we were in our teens, he used to let us drive the car; and when a bobby came in sight on his push-bike we had a perfectly developed act in which we would dive over into the back while Les would slide across and grab the wheel. If our headmaster wrote to him about one of us, he would regularly send us the letter, and he instinctively took our side in any hostilities with the school. Sometimes he would visit the school to preach, and he would enjoy hearing the topical jokes about the masters. He would regularly refer to them by the nicknames the boys used. We thought he was a pretty good father; in sermons he had a way of referring to children always as "little" children and they always said "Mummy", and this made the hard guys snigger; but mostly he didn't embarrass us saying precious things in front of the others. Once, preaching at the school at the time of one of the major battles of the Second World War, he began his sermon, "I have been instructed to conclude the sermon in time for you to hear the one o'clock news." He did.

We did lots of things together and, all told, had a lot of fun. We used to go to the flicks, the three of us, and sit in the one-and-nines in a draughty Odeon near the house in Finchley. Les never seemed to be able to follow the plot. "But Dick," he would complain in his loud whisper, "that chap with the hat: isn't he the chap who shot the girl?" Someone from behind would shush him and he would begin to snigger. "Well, I can't understand it at all," he would announce, as if this were a final judgment on Hollywood and all its works. Sometimes last thing at night he liked to go for a walk round the block and one of us would accompany him; we each carried a walking-stick. He would amuse himself by trying to flick or knock his partner's stick out of his hand; he had it down to a fine art and could sometimes send my stick flying across the street or even into the gardens of the houses; once, I recall, it flew up and came clattering down on somebody's roof. It must have surprised them. It was a good game; you won by scoring points.

Sometimes after the war the three of us would go down to the Dial House for a day or two, walking on the cliffs or watching the birds on the marshes, and then coming back to bacon and eggs or whatever simple fare could be provided from a can, and then gathering round a huge log-fire and lighting our pipes. Once when we returned in the evening the room seemed different. Things had been changed, and on the arm of one of the chairs was a copy of *Goodbye Mr. Chips*, with my grand-mother's spectacles lying on it. Les looked at it in horror: he had come immediately to a supernatural conclusion. "I'm not going to be haunted by that old dame!" he declared hoarsely. What had happened was that the cleaning woman was in the habit of letting herself into the house to have a nice quiet read, and she had made a practice of using the late Mrs. Triggs's spectacles which reposed in a bureau there. "Well, isn't that interesting?" said Les when it was all clarified, relieved that his mother-in-law was resting in peace.

Les was always glad to have us children about, and when we had been away and returned, his greetings were always royal and profuse. One spring we had all three been invited to dinner at Virginia Water. I was coming from Oxford, the others from London, and we were to meet; I hadn't seen Les for a week or so, that's all; but the meeting was like Stanley and Livingstone. I had arrived first and was sitting with the host, hearing the birds carolling it gently in the woods. Shortly the air was shattered by three long blasts on the Jaguar horn, and, thus heralded, in came Les, surging like a troop of cavalry and beaming like sixteen bishops in anticipation of our meeting. "Ah, *there* you are," he cried; and, filling the room with his magnificent euphoria, he continued to address me, contriving all in one grand sweeping movement to give up his topcoat, receive a glass of sherry, light a cigarette, shove away the cat, and occupy the best fireside chair before recognising the presence of his host and hostess.

We thought well of Les and he knew it, so that he put up with any amount of teasing and, indeed, relished it. There was a family rule that he wasn't to be teased about a sermon before he had preached it, and thus on the way to church we wouldn't even ask him his topic. But on the way home it was open season, and nothing entertained him more than the impersonations we used to render: we would catch his intonations and gestures and repeat the few words for which he had his own individual pronunciation; he would become weak with laughter, his shoulders rocking as he sat forward in the passenger seat of the car and tried to light a new cigarette from the butt of an old one. When he got the C.B.E. he asked us, "Why did I only get a C.B.E. when Alec Guinness got a knighthood?" "Well, Les, Sir Alec is funny *all* the time," we told him. (As a matter of fact he was never *quite* sure why he had got the honour at all. And he said that in the ceremony, when the Queen asked the man in front of him in the line what he had done, he (Les) became quite nervous that she would put the same question to him.) He used to say that when he was made President of the Methodist Conference one of us had cabled him from the States: "See Methodist Hymn Book, No. 503 for the hymn beginning, 'God moves in a mysterious way, His wonders to perform.'" Well, I think that was Arnold, actually, the son of H. B. Rattenbury, honouring *his* father; but it was the kind of thing either of us might have done, and Les would have liked it.

We were good for him, and we helped to hold at bay the loneliness that constantly threatened to overwhelm him. Perhaps it derived originally from childhood—the loneliness of a boy in a family with two elder sisters who didn't need him for their games, making his own world in the imagination, leaning over the back of a chair and commanding the applause of listening senates. It is clear from the early diaries and from later anecdotes and comments that he saw himself often enough as the lonely one or the odd man out—the young, unmarried minister in lodgings, the foreigner in Madras who couldn't speak Tamil, the teetotaller in the officers' mess, the reserved Christian among hearty pagans. With marriage, of course, his solitariness abated for a time; but even then most of the things he liked to do for fun my mother didn't care about: she didn't walk much (when she did, she didn't play the walking-stick game); she didn't care for theatres or movies or dinner parties; she didn't like to travel—I think for about twenty years she didn't travel further than she could go in her little Austin Seven, which she used to race up hill and down dale in Greater London unmercifully, thrashing it through the gears and making it roar like a Hawker Hurricane. They did little together, until late in their lives when Les had retired and they used to take the car in the afternoons down the narrow

Sussex lanes and stop to gather primroses, and in the evenings play Scrabble and watch Dixon of Dock Green.

His life until then was in any case essentially all work, and he did his work alone. Giving lectures and visiting churches all round the north of England, he travelled alone; most of his trips abroad were made quite alone. And any man whose life involves meditation or the use of the intellect is doomed or dedicated to physical loneliness, if not that loneliness of the spirit endured, so they say, by the saints or martyrs or anyone way out ahead. And of course he courted it: it became an early habit with him to withdraw from human society to commune with himself, or nature, or God; he didn't want to be a part of the noisy world and its necessary transactions; and it was for him a kind of sacrilege if my mother should invade the sanctity of his study and ask him to smell a newly-opened can of corned beef. He liked the idea of loneliness, and he liked the word: he saw Christ as lonely, and Paul, and all his heroes; and the great romantic poets had made loneliness a cult; in sermons he spoke of the lonely hills, the lonely stars, the lonely cry of the curlew, the lonely sea.

But he well recognised loneliness as a kind of horror, and he was unduly sensitive to the loneliness of people in London: young men and women in lodgings, come from the provinces or from abroad who were away from home for the first time; old people who were bereaved or crippled and cut off. He thought of the city often enough as alienating, an "unreal city" where "each man fixed his eyes before his feet". And thus in his ministry one finds an emphasis on community. The Christian church itself is so committed, but repeatedly in Les's sermons and in his informal remarks is the expression of his aim to make the Temple a place where people could meet and get to know one another: the Friday Fellowship was to be what its name denoted, and the Samaritan League too. The coffee sessions below the church after the evening services were designed to combat the loneliness of people in the city. From his own experience, he considered Sunday night the most lonely time of the week; on a wet winter Sunday night in the city, where was there a place for a person to go for friendship and conversation for an hour or so before toiling back in the tube to Golders Green or Streatham, to crouch over a gasfire in a basement lodging? Community was what a church owed its people, not as an attraction but as a central part of the Christian life.

Les knew what it was to be lonely, and if he courted loneliness as his needs indicated, he also dreaded it, no man more. In the company of his sons I believe he had the happiest times all through the 1940s: the days we did things together would have their dates circled in his diary—his way of indicating a bullishness in the happiness market. And so it was a great loss to him when, late in 1950 and early in the following year, both sons left England to live in the States. He felt this as a deprivation, I think. Throughout the fifties he mentioned it in letters—partly, no doubt, out of kindness but with meaning none the less. The loss, I think, was only really eased when, on the bright October day in 1960, he left London and all the stresses of that life and drove down to Bexhill with my mother to start a new kind of life with her, and on the way stopped near Haywards Heath to have a picnic lunch.

After the publication of *Psychology, Religion, and Healing*, Les appears to have become increasingly involved in discussion and controversy about spiritual healing. He had been for some years the chairman of the spiritual healing committee of the Methodist Conference. He was very satisfied to report at the Conference in 1952 that the British Medical Association now not only permitted but also encouraged doctors to work in co-operation with ministers. But at the same meeting, and again and again thereafter, he admonished his colleagues to oppose healing missions—the widely reported flamboyant ceremonies where the crippled threw away their crutches and the blind their white canes. His stand against these gave rise to a good deal of controversy and even animosity among people with psychic interests, who felt, I think, that he was their lost leader. He believed that the dangers of mass hysteria, under the influence of which a disability might be banished only to reappear later or to produce other more grave afflictions, outweighed the rather remote possibilities of permanent cures or beneficial effects. Over this matter he broke a lance publicly—but with the most civil terms on both sides—with Mr. Harry Edwards, the well-known psychic healer.

In the early fifties there seemed to be rather an impressive number of letters reporting that people the congregation had prayed for had been cured, or that they had progressed, or that their attitudes had been rescued from despair. The phenomenon has already been illustrated in these pages—the turning-point from death marked at seven o'clock on a Sunday evening, the perplexity of physicians, the joy and gratitude of parents or friends. In these days Les was becoming absorbed with the mystery of epilepsy and the possibility that this incurable affliction might be caused by demon possession. He derived great delight, as will appear, from the coincidence of the modern healing methods of science with ancient rituals and religious practices. At the same time he resented the strait-jacket of science, its dogmatism, and its insistence on its own procrustean terms. Writing to James Stewart, he says, "You know, Jim, we are too obsessed, most of us, by 'modern science' and we try to push the N.T. [New Testament] into its categories and it just won't go. You have had the courage to say so. Bless you and a thousand thanks." He himself had found that by suggestion under hypnosis he had been able to lengthen the period between attacks in epilepsy patients. One man had gone for two years, whereas before this treatment he had been suffering attacks three or four times a week. A girl he was treating had gone much longer. "In the privacy of my room," Les wrote, "with the patient hypnotised and unconscious before me, I have even used the old formula: 'In the name of Jesus Christ come out of him.'"

One of the most remarkable species of healing that began to engage Les during the early fifties was odic force, and one of the most rewarding human encounters was his meeting with Dr. Michael Ash. Late in 1951 the Bishop of Lichfield, who was a personal friend of the royal family and was concerned about the health of the King, received two letters which he shortly forwarded to Les. The first was from a Church of England clergyman in Bristol who described the miraculous effect that the sacrament of anointing had had on Sir Stafford Cripps who had lain at death's door with cancer and T.B.; it suggested that perhaps the needs of the King might be met in a similar way. The second was from the Bishop of Rochester who noted that the Queen had become interested in spiritual healing as practised by an American evangelist in England and suggested that, if she were interested in getting help of this kind for the King, the services of Dr. Michael Ash might be well employed "and with wonderful effect". The letter proceeded to point out that Dr. Ash was not in fact himself a spiritual healer but a qualified doctor, and it described briefly some of his methods. Rochester related that in his own knee, ailing for some time with fibrositis and rheumatism, one treatment had banished the pain, and he concluded with his belief that Ash's kind of treatment would bring the King great benefit.

In transmitting these letters, Lichfield said he proposed to pass on to the Queen the information about Michael Ash, but wished first to hear what Les thought about this kind of healing. Les then in some trepidation went to call upon Michael Ash in his consulting rooms in the West End. He stood on the step and wondered what he might say—"I've come to check you out for a bishop friend of mine"? Well, hardly. A flunkey appeared who asked if he had an appointment with Dr. Ash. Les was always shy before butlers and footmen and men in livery with silk hats and medals whom one never knew how much to tip. "No," he admitted, "but here is my card. Perhaps for a minute . . ." The morning suit glided off with the card in its hand. Then, in half a minute, a young black-haired man came bounding into the hall holding the card and declaring in a series of broken phrases he was delighted, it was a terrific surprise, etc., he had always wanted to meet him; he had heard him preach at Cambridge years ago; he had changed his whole life, etc; he must come right in and see what he was doing. Les was divested of hat and coat and ushered into the room where a naked woman lay prone on a couch. "This is Dr. Weatherhead," explained Michael Ash, perfunctorily. "Now run your hand down the spine about a half inch away from the skin." Les blinked and did so. "There! You can feel it?" He confirmed that he could: although he wasn't touching the spine, there was a tingling in his fingers as they passed over a certain point. "That's where the trauma is," said Michael.

The cause of the tingling was a leakage of psychic energy, which could be detected, as the presence of water can be detected by a water-diviner. Every element, Les learned, sends out radiesthetic waves, and as the emanation from the water meets the emanation from the diviner, a muscle-twitch results. Similarly, where there is pain in the body, psychic energy emanates and meets an emanation from the fingers, causing them to tingle. Michael Ash first thought of the leakage of body psychic energy only as a means of diagnosis, indicating the site of a trauma; he soon learned, however, that the act of holding the hand over the site of the leakage had itself a therapeutic effect. He found that he was not only stopping the leakage but in some cases "was able to feed energy into the patient's body at the site where it was leaking out. This became a routine method of treatment."

The name of the healing energy that emanates from the finger-tips is odic force. It was discovered first in the nineteenth century by Carl Reichenbach, a Stuttgart chemist, who seems also to have introduced to the world both creosote and paraffin. It is discharged through hands and through breath; it is retained in oil. Les was excited to learn of this healing power, the properties of which seemed to correlate so aptly with such ancient religious practices as the laying on of hands and the anointment with oil. It seemed also to be in harmony with an observation of Hippocrates in the fifth century B.C.: "It hath appeared oft while I have been soothing my patient, as if there were some strange property in my hands to pull and draw away from the afflicted parts aches and diverse impurities, by laying my hand upon the place, and by extending my fingers towards it." He was even more excited to realise that he possessed the power himself.

Not less amazing to him, and a very great pleasure, was his acquaintance with the young doctor, who combined his commitment to science and his religious beliefs, and brought both to play in his medical practice. He used conventional medical techniques and he used odic force, and at the same time he prayed to the saints to intercede.

> A case I remember very well, [he says] partly because I have not yet found time to fulfil a promise I made, was that of a little girl who was brought to me when she had only a short time to live. I detected an energy leak over her shoulder and traced it to a nerve in a spinal segment in the neck. The nerve also supplied the diaphragm and affected the liver; the child was as yellow as a guinea. I considered the case incurable, but corrected the position of the spine in the neck and made a prayer as I did so that if the child recovered I would go to Fatima, in Portugal, to say "Thank you." The child did recover, but I have yet to fulfil my promise.

He believed that power from the hand could trigger off a self-healing reflex action in the body. And once when he was out riding with his wife, she was thrown and dislocated her shoulder: "I used the power from my hands on her shoulder," he writes, "and was able to induce a very strong muscular reflex action. I watched with growing fascination as the automanipulative process took control of the operation. The work of automanipulation went on stage by stage, as though the body itself knew consciously and exactly when a period of rest was needed, until the shoulder had reset itself."

When Les first met Michael Ash he was working in London but returning at week-ends to his home in Cornwall in the far south-west of England. He told Les how he used to practise teletherapy, thinking about his patients, imagining their pathological appearances, as he travelled in the train. Once on such a journey he became conscious of the strong stench of cancer, and he learned later that at the instant he had sensed it, a patient had taken a turn for the better and the cancer had begun to leave his body. He had interests in all kinds of therapy, mostly related to radiation. He believed cancer came from smoking because of the radiation emitted by the metals in the smoke. He believed, what has now I believe been confirmed, that cancer arose, again from radiation, when there were metals in the water supply, as in areas where the ores of tin and lead contaminated the water in the valley. He invented a kind of sweet which was to counter the radioactive fall-out from bomb tests by preventing the absorption of minerals. And he invented a pill of vitamin C, kaolin, and charcoal which was to beat the breathalyser by breaking down the alcohol in the stomach.

Fortified with this charm, like Ulysses with the moly, he had a drinking contest with his wife on television and remained sober while she got drunk.

Les himself began to use odic force in the early fifties: the first patient on whom he used it was a woman who used to appear at the church hobbling stiffly between two canes, her feet scarcely in control, swinging out and falling erratically forward. Her sickness had been diagnosed as Friedrich's ataxia, for which no cure was known, and she had been instructed by her doctors that she would shortly be confined to bed. She was of unusual courage, however, and would not be put down by the dark clinical prognoses. Les first practised hypnotism on her, and the suggestion he gave her seemed to improve her morale. He kept up the treatments, widely spaced though they were; but the condition, while it did not seem to worsen, did not improve. Preparing for hypnosis on one occasion the patient complained of a sharp pain between the shoulder blades. Les passed his hands over her back without touching the skin, and there was a sudden convulsive movement of the muscles and the patient felt a most comforting warmth and complete relief from pain. She lay on the couch; Les found that wherever his hand moved a few inches from the body, the muscles twitched and the patient felt the same pleasant glow. He proceeded thus to work over ankles, knees, and hips, which became looser and more flexible under the emanations from the fingers. The patient returned periodically for more treatments, and muscles which had been stubbornly locked for twenty years were brought to respond easily to the will. The patient was amazed at her renewed abilities and at the speed with which they had been returned to her. She began to do all the exhausting things, determined to try out her new physique—she went abroad, sight-seeing and visiting all the galleries and museums; she went shopping; she drove a car through London; everything worked. Les used to call her his guinea-pig, since she was the first on whom he had used this new technique. She wrote, "The guinea-pig is very grateful but I must admit that I find odic force most uncanny. All these miraculous things happen and you don't actually come into contact with the body."

But the healer himself, what of him? The fifties were in general a bad time, and there were no miracles. It was in the fifties that he first began to get pains which were diagnosed as diverticulitis, a condition caused by the tension in his way of life which, insusceptible of cure, dogged him on and off right into his old age. Then he had to have an operation which, unsuccessful, had to be repeated. And these, painful enough in themselves, were followed by a skin condition which again plagued him from time to time for the rest of his life. The treatment was a nuisance: paint with this and that, take cortisone and ultra-violet light therapy. He resented all the time that had to be given to this rigmarole, in spite of which the specialist told him glibly, "Try not to think about your skin!" Then insomnia, "my old enemy", as he calls it, returned to disturb his peace by night, and anxiety haunted him by day. He sent me a clipping from the London *Star* which began, "From the clear face and hazel-flecked eyes of Dr. Leslie Weatherhead, physical and mental relaxation flows like a soothing stream", but he had written in the margin, "How easily a reporter can be deceived!"

Although the idea had been with him at least since the outbreak of war, I think it was around this time that he began to talk about his death as a translation to be looked forward to. At the end of a letter dictated to his secretary and typed, he has written in his own hand

> I feel moved to say without being morbid, that if anything happens to me don't mourn. I shall try to help you from the other side. I believe intensely in life after death. I have had a good innings. I've nothing more to say . . . I long to be rid of a body that aches and is in pain. Every appointment now is a burden and every service a strain and I am getting to be a neurotic old hypochondriac always worrying about my health. On the "other side" I should be gay again and free from pain and able to climb the equivalent of the Moffat hills.

This was how he felt and spoke about death in those years. Then later, aged seventy-five, when after peritonitis he came near to death, he described it in a letter to James Stewart as a "wonderful experience; as if one were in a field on a dull, grey day and then saw through a gate sunshine and happy people and incredible joy. I just LONGED to go through the gate. I prayed—as I've never prayed before—to die, but was hauled back *with immense reluctance* by antibiotics!! My doctor said 'I saved your life' and I said 'What for?' "

It was after the second operation that he began to receive messages from persons who were dead, through the automatic writing of Miss Geraldine Cummins. He had contributed a foreword to *Nurslings of Immortality*, a book by his old friend Raynor Johnson, and it was in connection with this that a message for him about his health came to Raynor from his deceased friend Ambrose Pratt, who transmitted it through the automatic writing of Miss Cummins. She then proceeded to transmit messages directly to Les from dead persons who were concerned about his health and later about his earthly and spiritual career. Over a period of ten years, she mailed scripts to him. Each began in the awkward childish hand of "Astor", her control—the opposite number on the other side—and then settled into a long slanting hand which joined together most of the words in the line. The early communications came from his father, Andrew, and his mother and Maxwell Telling, the physician with whom he had collaborated in Leeds. His father-in-law, Arthur Triggs, also came through. Later, after the publication of *The Christian Agnostic*, there were letters of admiration from Sir Oliver Lodge and F. W. H. Myers.

In the end, he was never quite certain whether the messages came in fact from the afterlife or whether they came, miraculously enough even so, telepathically from his own conscious or subconscious preoccupations. But as the early scripts came to him he was strongly impressed: he wrote to Miss Cummins, "You have enormously increased my faith in life after death." In a communication in 1956, Maxwell Telling observed that "The ills that arise from too rapidly consuming the grey matter of the brain cannot be overestimated." Les found precisely the same sentence in a textbook on biochemistry. He thought perhaps Telling had read it in his lifetime: "If you have never seen this book", he wrote to Miss Cummins, "the evidence that the message is from Telling seems to me valuable, or else how could these words be quoted word for word from a textbook published at St. Louis in America in 1833?" Telling recommended potassium phosphate to Les, who told Miss Cummins he was taking a lot of it.

Another curious message came to him in the form of a letter from a soul in Purgatory, transmitted in her usual way by Geraldine Cummins, who later recalled her sense of "deep unhappiness" as she obtained this part of the script. The writer of the letter told the following story: he had written a book called *Cancer*, for which dread disease, having devoted years of research, he had found both a cure and a means of prevention. After making his discoveries he had set out to force them upon the attention of the medical world. However, becoming involved in a sexual scandal, he had ruined his chances of publicising his work. He might have gone over to America to make a new start, but he had lacked the courage, he said, and he had taken his own life. Part of the purgatory he now endured had been to learn that cancer research had made no progress and indeed was on the wrong lines. At the end of his letter, he begged Les to call attention to his work: he wanted no fame; others might have the credit. All he wanted was to be liberated from the obsession of his cowardly death and his "betrayal of armies of cancer victims". The cure he had found involved potassium.

In 1964 a message came from Arthur Triggs which told Les that he and my mother would both live till beyond their eightieth years. She would die first. Death in any case was not to be feared: "Your long well-spent lives," wrote "Astor", "and the deep love of certain souls who have gone before you and during your last hours will be near you, waiting and watching for you, make of death for you both merely a dreamless sleep from which you wake to find that you have come home. So welcome death as your best friend for you wake in your family circle and are greeted by your beloved dead who at last become wholly alive to you again." His ghostly father-in-law added also that he had the impression that Les was composing a book which would make a profound impression and "help provide a true simple religion for daily life. It will needless to say therefore be attacked by the traditionalists and back numbers."

After the destruction of the church in 1941 when Les, with Alice Head and Albert Clare, began the task of rebuilding, there was for many years little action they could take except to raise money, and this they began to do. Most of it came in from appeals of one kind or another, and members of the church contrived to round up funds from all kinds of miscellaneous efforts, of which concerts given by the church orchestra, admission sixpence, and jumble sales were the more conventional. Some offered to do embroidery with profit for the church, others to make teacloths, repair radios, or baby-sit; people gave jewelry or trinkets from their attics, or foreign coins from trips abroad in happier times. Frank Salisbury, the painter, offered to paint Les's portrait and donate it to the church if five hundred pounds could be contributed to the Rebuilding Fund. It was not the first occasion on which Les had sat for Frank Salisbury: on an earlier occasion the painter had been commissioned to do a portrait of Bishop Asbury, and lacking a live model he had asked Les to put on a wig and robes.

Among some, the zeal for the church ran high, and Les had to ask that lotteries and other games of chance and gambling should not be resorted to for raising rebuilding money. He was the victim of zeal from another quarter: in the fifties he served on the Brains Trust, the popular B.B.C. feature in which questions were submitted to a panel of distinguished men and women and discussed over the air. Unknown to Les but in his name, one member of the church sent round to all the fellow members of the programme an appeal for money for the church. When he learned of it much later, Les burned with shame and wondered what Julian Huxley, Sybil Thorndike, and Jacob Bronowski would think of

With Mrs. Weatherhead and Gertie Hutchinson at Bexhill.

In retirement, in the garden at Bexhill.

him; but at least he now understood why he had been dropped peremptorily from the programme for reasons that were not apparent at the time.

Most of the rebuilding money came from direct appeals by Les to the congregation of the church. Or fellow ministers in the provinces would invite him to preach and would then share the collection with the City Temple. Money came from afar—especially after the war was over, when Les had begun to make his journeys. By this time the church officers had started to formulate plans. There were some radical decisions that had to be made, of which the most serious was whether to build on the old site in the City or whether to seek a new one in the West End. They considered also the idea of settling down permanently with the Presbyterians at Marylebone. There were many hours of anxious discussion and debate over these questions before finally the old site was selected and the architects appointed. All the moves in the game were reported blow by blow by Les in the church magazine and often enough repeated in the religious press—the various decisions made and all the reasoning; the negotiations with the War Damage Commission; the request for building permits; the testing of the calcined stones for their strength with a view to re-using the old fabric; the shoring up of the walls and the tower as these became dangerous.

In February 1954 the Lord Mayor of London gave a lunch at the Mansion House to launch an appeal for two hundred and fifty thousand pounds. By this time Les had made his plans to visit the States, to preach and lecture around the country, telling his story of the destruction, and letting it be known that he needed funds. When he sailed on the *Queen Elizabeth* in March, the figure needed to rebuild, the "gap" as Les was pleased to call it, was at a hundred and seventy-two thousand pounds. He had no great confidence in the outcome of his journey; he was not pessimistic, but far from optimistic: he was acquainted with many Americans who were outgoing people of incredible generosity, but he was setting out to raise a colossal sum, and he knew of one English minister who had travelled to the States and across the Continent and had failed to raise even enough to cover his expenses. For personal reasons, Les didn't very much want to travel, to disrupt his work in London, to leave home. And he feared not a little for his health in the strenuous exertions that were to be demanded of him. Some of his fears materialised; others were gloriously dismissed.

He was met on the dock by his old friend Lynn Harold Hough and O. Dickenson Street, who made him his guest while in New York at the University Club on Fifth Avenue. He stepped almost immediately into the whirlpool of luncheon addresses, lectures, and services. And then almost immediately the major objective of his journey was practically fulfilled in a stroke, and the burden he had borne in his mind for over a decade was lifted.

He preached for Ralph Sockman at Christ Church Methodist and at the (Congregational) Broadway Tabernacle. At the request of Helen Keller he read the whole of his book, *The Will of God*, into a recording machine to make it available for the blind. At one luncheon Bishop Pike was in the chair and when he praised Les and his work with some extravagance, John D. Rockefeller, who was the host, whispered to him, "That's what we all think." Rockefeller invited him to lunch at his place, and there Les described his mission—the needs of the church and his plans on this present trip. Rockefeller was not a little shocked at the extent of the preaching and speaking engagements scheduled, and begged him to cancel an appreciable number of them. Further, he insisted on sending a tour manager along with him to deal with details; and thus Mr. Harry Fish, an economist from the Rockefeller Foundation, went with Les on his travels and managed the reservations on planes and in hotels, ordered taxis, held the tickets and the baggage checks, looked after all the accounts, and headed off ambitious reporters. He was a fine guy: a widower in his sixties, he was good natured and, as it proved, more than somewhat long-suffering. His assignment, to accompany Les, exposed him to a good deal more theology than his system was accustomed to and more hymns than were usual in his diet. But he bore up. Only occasionally, after a service or a lecture, anticipating one of those receptions where you get coffee and angel-food cake and have to take nuts with a spoon, he would whisper to Les, "I think I'll just slip out and sample the local beer."

A few days after his lunch with Rockefeller, Les received the following letter from him:

> Dear Dr. Weatherhead:
> Toward the fund of $474,000, which is the minimum amount required to rebuild the City Temple Church, London, in addition to the funds already secured or pledged, I am glad to contribute $300,000 to be paid when and as you may request within the next six months. Toward any balance of the fund of $174,000 which you do not succeed in raising during your stay in this country, I will contribute up to an additional $100,000.
>
> Very sincerely,
> John D. Rockefeller, Jr.

Thus was lifted the major part of the financial burden.

Most of the journey remained, however. Les left New York at the beginning of April for Philadelphia, where he preached twice, then to Washington for a round of engagements, one of which was the opening of the Senate, habitually effected with a prayer, normally offered by the Senate Chaplain. He met Nixon, then vice-president, and a number of senators. At Chicago he preached at Northwestern University, addressed the Sunday Evening Club, and played a round of golf. He went to Cleveland, Detroit, St. Louis, and then at the end of April down to Wichita Falls, Texas, to deliver the Perkins lectures, a series endowed by an oil man.

In Wichita Falls, there were eight lectures scheduled from Monday to Thursday, and Les was to preach twice on the preceding Sunday. After the morning service, however, he began to feel the acute effects of the strain he had been imposing upon himself now for six weeks. He had, of course, accepted more invitations than his strength could underwrite, and often, in addition to what he had agreed to do, he would arrive at a city to find that three or four minor engagements had been slipped into the schedule. Important people of the church wanted him at luncheons and tea-receptions. He was instructed that he couldn't disappoint Mrs. Thingummy, for she had put up new drapes in her living-room for the occasion and had had the Wallmaster people come in. Also, in forming his schedule months earlier in England, Les had accepted speaking engagements without allowing adequate time for travel and recuperation therefrom. He had given no thought to the bodily taxation incurred by changes in temperature—he flew, for instance, from Detroit in winter cold to St. Louis in warm humidity, with no time allowed for acclimatisation. He suffered from air-sickness on occasions. The night's rest, short enough most often, was frequently eroded by the inevitable delays of air and road travel.

After the morning service in Wichita Falls, he excused himself from lunch and lay on his hotel bed wondering what on earth he could do about the evening service and the eight lectures he was supposed to be giving, beginning on Monday. He wrote later, "It was then that God answered my despairing cry for succour. I had preached during the tour on 'the God who will meet us at every corner', and here I felt I was being challenged to put to the test the faith I had so often preached. How often I have preached that God comes to us through others." He explained how a doctor was called, who, deducing that his endocrine system had collapsed, gave him injections and massage and enabled him to get through the week. "His help," he concluded, "seemed to me an answer to prayer."

He had to prune away parts of his schedule, however. He went next to

Memphis, where Dick, his elder son, worked in a mental hospital. Then, accompanied by Dick, he went to New Orleans, forgetting his notes and lengthening the motor ride, and then out west to Berkeley and Tacoma, Washington. Here he delivered a commencement address and picked up a D.Litt., a degree which only three other people had been given—Lord Halifax, Basil Matthews and Lynn Harold Hough. From Tacoma, my wife and I took him on a car trip through the ancient rain forests of the Olympic Peninsula. We stayed in motels and cabins amid the extraordinary scenery of that territory and fixed our own food. Les did his share of the housework. Once I remember him looking up from his hands plunged in the dishwater. "When I stayed in New York," he announced, putting on the mock pompous tone, "as a guest in the University Club, I'd have you know I was provided with a valet!" In one cabin we stayed in he left his card with a note: "I think you should know: there are mice in this motel." It was some time since he had been able to spare the time or the mental energy to consider the minor imperfections of this world. He lay in the sun on the rocky beach at La Push, an Indian village in the most westerly part of the continent, and watched the small waves of the Pacific running almost soundlessly up among the round stones, and he began to recover his vitality.

He went back to New York to preach at Riverside Church and after the service he was driven out to spend the day with the Rockefellers in their country place in the Pocantico Hills. In the evening they sat on the terrace while the sun went down behind the hills on the west bank of the Hudson River. Les wasn't watching the sun, because his host was working on the figures he had given him, showing what had already been given or pledged to the City Temple and what still remained to be found. Mrs. Rockefeller called to them to come and look at the sunset. But Les, though uncommonly fond of sunsets, excused himself, since his host was in the process of giving him another 100,000 dollars.

"Come on," said Mrs. Rockefeller. "Come and look at this sunset and I'll give you a hundred thousand too!" Les, at that, complied: he was a connoisseur of sunsets and, he reported later, it was a singularly fine spectacle. But at the actual moment—the moment in which he realised that the whole financial problem was settled, that the church could be opened free of debt, that he could go home with the good news that might be shouted across a street—in that moment he was too moved with gratitude to say anything at all. Rockefeller filled the awkward pause, declaring that nothing that he could do would repay him for everything his books had meant. "We are both so happy," he wrote the following day, "to have this opportunity of helping in the completion of your enterprise and in so doing to show our deep appreciation of what you have been and are to us spiritually."

The Irish chauffeur who drove him back to the city prattled away, but Les heard not a word.

> My heart was too full [he wrote] of exultant gladness that our burden had vanished, that I had such wonderful news to bring home . . . and that God had put it into the hearts of His dedicated servants, especially one of His humblest servants, to help set up the City Temple again in the heart of the great city of London, rising from the ruins of war to speak to the world through many centuries of those things which cannot be destroyed; the things by which alone men truly live, the things which shall bring peace to all the nations and gladness to the hearts of men.

So Les returned to London with the half million dollars and more, and the building plans went ahead, and the building itself. And then at last there came the day in 1958, August 17th, when he preached in the new building on "The Gladness of Returning Home", quoting first the text, "I was glad when they said unto me 'Let us go into the house of the Lord.'" And then this: "Remember in gratitude the debt you owe to Almighty God, Who has permitted you to efface the last trace of the ravages of the great fire, and to assemble once more for prayer and praise after so many years, on this spot consecrated by the devotions of generations"—words of the Bishop of London in the new St. Paul's after the Great Fire, spoken in 1697.

II

Les returned from the States in time to attend the Methodist Conference, meeting that year in London, at which he was elected President-Designate, to serve from 1955 to 1956. The election cannot have come as a complete surprise, since in previous years his name had appeared on the ballots and he had won votes. But he was impressed with the honour paid him, particularly since he was not then actually serving Methodism—not technically, though he preached continually in its churches in England, and half his ministry in Australia and nearly all his appointments in the States were under Methodist auspices. One Methodist dignitary described him at the Conference as the Prodigal Son of Methodism; and Les, following up the allusion, said he was sure the City Temple people would graciously accept their role as the pigs and the harlots of the far country. The Temple freed him to fulfil the duties of the office and gave him a new Rover to take him about. But in the interim, for the twelve months between the election and the year of office, he dreaded the prospect of the presidency and its exhausting demands; and in bad moments, lying awake in the small hours, for example, when his problems always assumed their most monstrous shapes and dimensions, he determined to flick it in. But he held up; in July of 1955 he duly journeyed up to Manchester to be inaugurated. He had asked advice of his son, Dick, telephoning him in the States; and Dick had said, "Have a bash," a phrase recommending positive action.

The Methodist Conference, like the conventions of other callings or like the annual political meeting, consists of representatives from the periphery who have come for a week to the centre of things to shape or hear policies, to second a motion, to see old friends and have a good time, to preside over a working session, or to read a committee report. Most of the men were in dark suits and clerical collars. But some of the delegates were women; some of the men had brought wives and sisters to the Conference; and among the suits there were flowered hats and summer dresses in the auditorium of the Albert Hall, and there were some Indian and African visitors in their coloured costumes. Outside the main hall in the adjoining quarters there were encounters and reunions and cups of tea and gossip. The weather was hot.

Les got through the business with marvellous celerity: meeting-minded people, I expect, would have complained that he railroaded things through. Certainly he was a bit cavalier on occasions about procedure—I don't think he had ever studied it or cared much about it. Eric Baker, the secretary, who knew all about procedure, had to bring him into line from time to time; but once he whispered to Les, "Don't look over to the right: there's a chap there trying to catch your eye, and he'll speak for half an hour." Les kept things rolling. He was witty and charming through the long sessions, and no one got rattled; motions were made and seconded and voted on; there were the usual "in" jokes and one or two bits of badinage with Donald Soper—a regular feature of any meeting at which both of them were present, and the occasional ripples of laughter gave delegates a chance to shift in their seats and unfold their legs, and the ladies to fan themselves with their programmes. The lunch recesses came on time, and tea-breaks were observed with British scrupulousness.

In his presidential address, Les was very specific: no general hortatory rhetoric about bombs, and slums, and hungering Asians: he got right on to the immediate thing—the inefficiency of Methodism. There were, first of all, towns where four or five churches were maintained with all the overhead expenses, while all the Methodist worshippers on a Sunday could be accommodated in one building. He would close four out of five churches, he said, and concentrate the resources and the ministers in one place, which would become a "thriving vital centre of spiritual life". What would the communists do, he enquired, if they had premises and a paid agent in every community in the kingdom? They would proselytise the whole country in ten years. Next he called for tolerance; and here he specifically cited the reception in England of Billy Graham. He had supported Billy Graham's crusade in England and had countered hostile criticism which had thrown repeated punches at Graham's strategy of arousing emotion in the crowd. As we have seen, he defended Billy Graham on this score. As to Graham's theology, Les did not endorse it all: but men are changed by news, not views, he declared. Ministers ought to be grateful that Graham was helping to fill their churches. "We can teach people theology," he had said, "when we have got somebody to teach." Graham's crusade was a gateway: it was not a substitute for the church, but he thanked God for such a sincere and humble servant as Billy Graham who could do, he said, what he himself could not do and win people he could not touch. Thirdly, he asked for dedication. "What could God do with Methodism," he concluded, "if, at His feet, we began again; if we repented and really believed the Gospel!"

"Methodism anticipates a memorable year!" proclaimed the *Ipswich Evening Star* in headlines; and properly enough: Les had taken over! Later he had some other sensible housekeeping suggestions for Methodism:

raise ministers' salaries by 50 per cent, cut down the size of committees, on which, he said, some were vegetating who had "never had one bright or constructive suggestion to make in ten years". He would withdraw Methodist ministers from Scotland, and he would unite the church with Congregationalists and Baptists. All reasonable ideas, rash only in their simplicity and in Les's studied disregard of awkward working details. The Prodigal Son of Methodism had moved in and, replete with the fatted veal, was standing on the hearth telling them how to run the farm.

Conference week was tough, though. In addition to the sessions over which he had to preside, Les had to speak at lunch meetings and, often enough, preach in the evenings. In between times he interviewed occasional patients, who had come to Manchester in the hope of seeing him; he read his letters, and he answered phone calls that had come to the hotel in his absence. The hot weather didn't stop; and for much of the time he was in pain, thanks to old afflictions which flared up again as he drew on his resources. All in all, the ten days of the conference were quite an ordeal. The night before he came back to London he telephoned me to get some theatre seats. I met the train; we took a taxi to the Liberal Club, had a quick supper and then on to the theatre where we had seats on the front row of the orchestra stalls. It was a musical and very funny; Les slumped back in the seat and laughed through the whole show—two hours of therapy, while the mind laid by its troubles.

The year following, his year in office, was pretty gruelling. As president he was expected to visit every Methodist district in the kingdom, and three Sundays out of four these duties took him away from London. Ministers from all over the country wrote to him claiming, sometimes in error, that "the President always visits *my* church". In the press photographs of that year we see him in all parts of England, signing the visitors' book, shaking hands with the local minister, or standing in a posed group with the minister and the mayor wearing his chain, and one or two other worthy citizens of the borough. Always the smile; but here and there, in the unposed shots and unguarded moments, there are signs of strain and weariness. And no wonder: everywhere he went the local people wanted to get the best out of him: always there were crowded churches and crowded overflow meetings in church halls. Regularly people were turned away. People wanted to meet him and tell him about themselves; and if they had known him earlier, they wanted to show him to their friends. The kind of programme he filled on some weekends is burdensome even in the reporting: one Saturday night in Birmingham, for instance, he spoke to 2,000 people; then he spoke to another 300 in an overflow hall and then to 1,500 in another part of the city. The following morning, Sunday, he preached in Leamington to 1,300 people and then in the evening in Coventry to another 1,500. They were exhausting days; but he held before him the image of John Wesley, barn-storming England on horseback, indefatigable, carrying the Word. He preached on many of his usual topics in his usual vein: he told people that Christianity was a relationship and didn't depend on belief in creeds; he described his concept of the will of God; he spoke of forgiveness of sins; he reiterated his admiration for Billy Graham; he called for the unity of churches. He touched on a number of national issues—capital punishment, the prospective marriage of Princess Margaret to Peter Townsend; the government's Premium Bonds; its Cyprus policy. Whatever he said, the press sat up and took notice, and it was interesting to notice which remarks gained headlines or bold type-face. "Fun Making at Spinsters a Cruel Disgrace", announced the *Scots Pictorial* in inch-high type; "Mothers Can Help Brides-to-Be", said the *News Chronicle* on a day presumably when sensations were in short supply. When Les complained that the valiant young men who had won the Battle of Britain were settling down to a "butterfly whirl of meaningless inanities and giggling away their days", the *News Chronicle* had a headline over four columns which read, "Weatherhead Lashes Out at Butterflies", creating the image of rather a grotesque pastime for a sixty-two-year-old cleric! Occasionally his enthusiasm seems to have carried him a little beyond realism, and under the headline "Film Star Pin-Ups are 'a Major Misfortune'" he is reported as recommending that instead of much-married film stars, young people should use the photographs of missionaries for their pin-ups—a suggestion which seems to miss the point rather widely!

The lectures and services were taxing. He would leave home on a Saturday afternoon, three weeks out of four. He told me how he used to get a sick, sinking feeling when he saw the taxi draw up in front of the house to take him to his train. Along with the burden of speaking and preaching to strange audiences there was also, often enough, the awkwardness of staying in the houses of strange hosts and hostesses. As an insomniac, he missed the familiar things and the comforts of home that he couldn't expect to be duplicated elsewhere, even in the lavish guest-rooms of the rich or the rooms of old friends. Occasionally the physical discomforts distressed him—unaired beds, rattling windows, dogs that howled, and lavatory cisterns that roared in the night. One feature of other people's houses that regularly annoyed him was clocks. They chimed and kept him awake; even their ticking disturbed him, especially when they were large clocks and there was a period of breathless suspense between the tick and the tock. Against rattling windows he carried wedges or a few golf tees; against dogs and toilets he was powerless; but against clocks he thought nothing of taking serious counter-measures; he was a master at silencing them. Staying at Douglas Britten's, he slipped a Swan Vestas match-box under the foot of the grandfather clock so that the pendulum rested at one end of its swing. In the morning Doug told him that that clock had been his grandfather's and that it had been running for a hundred years. "Time it had a rest!" said Les, impenitent. In another place he stayed he was less fortunate: he crept silently downstairs in the unfamiliar darkness in order to stop the clock, quite innocent of the fact that the foot of the stairs was the habitual sleeping quarters of a large black dog. When Les put his bare foot firmly on the rib-cage of this animal, it sprang up and began to yell and bark until the whole household turned out to see what was happening. When he stayed in Leicester with Fred Carmichael, there was a little silver clock that told the quarters in its little silver chime; and when it resisted all reasonable methods of discouragement, Les hit on the scheme of burying it deep in a drawer full of extra blankets, so that its voice was heard no more. The Carmichaels discovered it months later in the autumn when the cold weather had come on; in the meantime they thought he had nicked it!

At the end of his conference year he went to hospital for the operations. The body is a tyrannical servant; but, as so often, it waited upon its no less tyrannical master's convenience before it handed in its non-negotiable demands. While in hospital, Les planned *A Private House of Prayer*.

The operation was a failure and had to be done again; and after the second ordeal his recovery was so slow that the doctor recommended him to go away for a holiday, into the sun if possible. Thus in February of 1957, the month he disliked most in the calendar, he took off with Ernest Appleyard for the West Indies. They stayed in a place right on the beach where, in spite of the roar of the waves, Les slept without drugs for the first time since he had taken up the presidency.

On the first Sunday of October 1957, he was greeted by a church packed with people. His assistant minister read the lessons, and then, remarking that this was their minister's twenty-first anniversary, he asked the huge congregation to sing the doxology. Les was moved by the voluminous vocal tribute. He was a bit embarrassed, probably—glad that the people wanted to express gratitude in this swelling music but wishing slightly that they wouldn't. He felt that the church was strong, and he had some satisfaction in this strength without particularly taking any credit for it.

He was himself, I suppose, at sixty-two, intellectually and spiritually at the height of his powers; if religious matters can be measured in the words of the world, he was a success. Through huge congregations and through broadcast services which were relayed to all parts of the globe, his voice was reaching more people than ever before. He was frequently on television. From all over the world he received messages of gratitude—"I heard you broadcast . . ."; "I read your book . . ."; "You have changed my life . . ."—letters which came in repeatedly day after day, and after a broadcast came in in shoals. The church clinic was thriving and men and women were leaving it cured. When, each October, he designated a Sunday as Gift Day and asked for a thousand pounds, he got over a thousand pounds. Repeatedly visitors to the church wrote to him to record their gratitude at finding friends and solace for loneliness. The whole outfit was thriving.

For himself, though, he increasingly felt that physically he had crossed the watershed and was declining. He found that he wasn't much interested any more in new proposals in the church: they didn't come to him as a welcome challenge but as merely another item on a committee agenda, a bit of a nuisance and a bore. His strength was on the wane. The winters for him grew longer. He used to parody Keats's "Ode to Autumn": the "season of mists and uselessness", he says in a letter one October; "I feel useless till next May about." Some winters around this time he escaped, going to Madeira with his friends Norman and Hansie French.

In the middle and late fifties things were running down for him. His ailments, the skin rash and the diverticulitis, afflicted him in bouts. Neither could be cured. The diverticulitis confined him to a diet: he could eat no fried foods, and he had to have all vegetables crushed through a fine strainer, a procedure that involved annoying complications when eating out in restaurants or as a guest in other people's houses. Then, also, in these days old friends died: he might look down from the pulpit and see the pews where they had sat, occupied now by strangers. In his letters there was regularly a note of the sad passing of time, of health, of life: "Poor old Hopkins is gone. Bill Sangster will never preach again."

His letters register another thing more emphatically than they had done earlier—his sense of the dullness of his days. He had had, in five years, the Australian trip, the American trip, and the Presidency of the Conference; and now with these exhausting efforts and packed days behind him it just seemed that there was nothing much going on. Part of the trouble was, obviously, that until he grew older and retired to Sussex and accepted a quieter regimen, he expected too much of life. His ordinary dull days were not more dull than the lives of millions, but he describes them as if there were a vital element missing. At St. Margaret's Bay in the summer of 1957 he writes ironically of the time he and my mother were having: "When we want a bit of fierce excitement we go into Dover to buy buttons or something and occasionally we go mad and go into Deal for a kipper . . . We go to bed at 9.30. It rains most of the time." Well, it was a change from flying to Tasmania at night and sighting the Southern Cross, or lunching with the Rockefellers, or commanding a spellbound crowd of thousands. But for most of us, most of the time, life is a noiseless procession of such small events, and holidays are a noiseless procession of smaller events in another less comfortable place where buying buttons and kippers constitutes a reasonable and adequate diversion. But, as we have seen, Les always had a rather Platonic contempt for the texture of living. He was a man walking in the city who thought of mountains. He didn't care to study the pavements.

If he found the Dial House dull, he found the weekly London routines even duller. All the activities which twenty years ago he had found engaging now failed to stir him—the letters of people needing help, the telephone calls, the day-to-day details that cropped up in the running of the church, church meetings, committees. Even the making of two sermons weekly sometimes presented itself as a load to bear rather than the glorious opportunity it had been formerly. In his spare time he found little to divert him. My mother was perpetually and chronically busy, Margaret was away in Oxford, the boys were in the States, his closer friends were in the provinces; there was no one to play golf with, but sometimes on Saturday afternoons if it wasn't raining he would go up to the park and run up the touchline of the borough football ground and tell the wing forwards when to shoot.

There were no outward signs. His work proceeded; his preaching was never finer; his engagement in the polemics of the church and the nation was apparently as spirited as ever. In 1958 he spoke up firmly on the side of peace when a general debate was opened concerning the alternative threats of nuclear war or communism. In this dilemma Les declared he would accept communism rather than the atom bomb. For once in these matters he spoke on the same side of the question as Donald Soper: he had shifted from the position he took in 1939 when he supported the war. But in fact it was not he but the whole panorama of possibilities that had shifted: he still used his familiar argument: communism now, like war in the thirties, was the lesser of two evils. In the same year he took a leading part in the Methodist Conference initiating a healing diploma for ministerial candidates. He did not apparently lose his familiar steam in these years; but what he did, he did at greater cost than before; and his activities left him a little more desiccated than they had in earlier days.

He had talked about retirement. In the early fifties of the century, when he himself was in his late fifties, he had mentioned that he wasn't going to go on for ever at the Temple. His father had died early in his retirement, and he hadn't had time to enjoy the leisure he had earned. Les, I believe, had this in mind. He certainly had it in mind to resign before anybody wanted him to resign. I suspect that in those days, as he became conscious of the fatigue that had begun to hang on his shoulders, he must have remembered quite often the words of F. W. Norwood at the inaugural ceremonies twenty years before: "London will squeeze a man dry . . ." and "when they have drained you dry, they may begin to say that you really have not got much to say after all." I think also that Les must have realised that he would need to be freed from routine duties of he were going to write the book that, over the years, had gradually been taking shape in his mind.

After the year of his presidency was over, he probably had some feeling also of the rhythm of his own career—the white-faced lad, giving his first sermon in the farmlands of Leicestershire, who had picked away the cord around the pulpit cushion in his nervousness, had become about the best-known preacher in Christendom; the young missionary, once lonely and unknown in India, had carried or broadcast his voice around the world; the novice who wrote a book and was charged with heresy, now according to a reader in the British Museum who had counted them, had fifty-five titles below his name; the young rebel who had laughed at bishops in their braces had become himself the top brass of the Methodist church. In the evaluation of the world, he had nowhere else to go.

There was, however, one more thing: the return of the church to its site on Holborn in the City. Since he had returned from the States the rebuilding of the Temple had proceeded apace. In the autumn of 1957 the shell of the building was complete. By February of the next year it was possible to announce that the Queen Mother would attend a service of rededication in October. By October, the building had been in use for two months: the people had returned home. In his homecoming sermon Les recalled the seventeen years' exile and some of its particular woes, above all those during the war, the death of kin and friends, and—though minor, appreciable—the weariness and the disruption occasioned by the church's eight moves about the city. Now, thanks to God, thanks to donors and workers, architects, builders, craftsmen, technicians, to those who had prayed and planned, they were home.

The building combined what was traditional about the old church with the innovations which time demanded. It had traditionally been a preaching centre and its so-designated "great white pulpit" donated by the City

Fathers had stretched across the end of the building. Now in that place was an apse, bathed in blue light, a shallow bay flanked on either side by forty-foot columns of cedar of Lebanon, enclosing a twelve-foot cross in cedar, which stood out from the wall of the apse, hanging, seemingly, on its own shadow; above it, in a stained-glass window, a dove wreathed in passion flowers; below it the communion table with flowers either side, and, spreading outwards, the crescent-shaped choir stalls, and behind them the ranges of the 3,500 pipes through which the organ blew its superb notes. The pulpit in brilliant white stone with crimson cushions was off to one side, approached by an invisible stairway and door. In the foreground on each side were stalls for the clergy.

The day the Queen Mother came, there were carpeting and a red and white awning over the pavement. Over the entrance the Stars and Stripes flew alongside the Union Jack in recognition of the American help which had been received. A procession of fifty clergy and laity brought up by Les and the Queen Mother, the city officials and the Mayor, moved slowly down the central aisle. Les was wearing his scarlet D.D. gown and purple hood, the Queen wore a hyacinth blue dress in Watteau silk, the Lord Mayor was in black and gold robes. The organist played the Toccata and Fugue in D minor; there was the National Anthem, hymns, prayers, readings and a short sermon. Les spoke briefly on his ambitions for the church: in conclusion he said, "Let the crowning glory of the City Temple be that it is a place redolent with love, full of the Unseen Presence, where the humblest worshipper shall say, 'Surely the Lord is in this place, and *I knew it*,' a place where the lowliest seeker shall find his way to the feet of Christ." Then the Queen Mother unveiled a plaque and made a short speech. The architects were presented and the church councillors. And then, in a brief moment the climax of weeks of anxious questioning, such as "Les, I have nothing to wear! What shall I say to her? Am I supposed to curtsey? Oh, Les, what if I should fall over? Whatever shall I wear?" my mother, without fainting or falling, said, "How d'you do, Madam," and the Queen Mother smiled. As she left the building someone less inhibited by protocol yelled, "Good old Mum!"

It was a great day. For Les, it was the day he had been awaiting for seventeen years. It was, he said to my mother, the greatest day of his life. "What about our wedding day?" she enquired, cooling off the enthusiasm a bit.

Miscellaneous honours came to him in these days. The church members had decided to have a bust made of him, which might stand in the entrance opposite that of Joseph Parker. It was sculpted by Kostek Wojnarowski (whom Les privately referred to as "Wodge"). The sittings were held in the study at home where the sculptor would come with his wet clay, slapping it and pressing it, arguing with his model about modern art and modern poetry. "I like to know what a chap *means* in a poem," Les declared one day when the sculptor was praising Dylan Thomas—the old Les, wanting a horse to look like a horse. He studied the clay head growing on its stand: "Well," he said, "I suppose it looks a *bit* like me." It does, though the upper lip looks as if it is trying to say "Prunes" without any help from the lower. There was a little ceremony at the church when it was unveiled, and Wojnarowski, a Pole who had had bad times, said that being in Les's company had recovered for him a sense of peace that he had not had since the war began. A compliment, perhaps, for a church supper, but felt, without doubt, and strikingly consonant with the experiences of many others with whom Les came casually into contact.

An honour of a different kind came to him when, in a small group, he was invited to a lunch at Buckingham Palace, where he was delighted to have a few minutes' private talk with the Queen. He liked her very much: he had seen her before the war at the garden parties when she was a child, and now he took much pleasure in recording that she was a beautiful young woman, natural and modest in her ways. He liked Philip too; and again it was the lack of pomposity that appealed to him; and he recorded how, when one of the fellow guests had inadvertently shot a small round potato off his plate and over to the Duke, the latter had crooked his finger behind it and asked the man if he wanted it back—an easy enough shot, no doubt, for a polo player.

In the New Year Honours list in 1959 Les became a Commander, Order of the British Empire, and went to the Palace for the inauguration, along with Uffa Fox, the yacht designer, Michael Somes, the dancer, an artist and a general, and the captain of a Union Castle liner. In the same list Alec Guinness received a knighthood, Rebecca West a D.B.E.; Kenneth Clark became a Companion of Honour and Stirling Moss got the O.B.E. One or two of the religious papers urged that a life peerage be awarded Les, and a small movement to this end was started by a friend. It would have meant much to him, I believe; he honoured such honours, and later when he heard on the morning news of the peerage awarded Donald Soper, he was the first to call up and congratulate him. But he was pleased enough with the C.B.E.: there were only six such honours among Methodist ministers, and the other five had been awarded on account of military service to men who were serving as chaplains; his had come, he understood, on the Prime Minister's list.

The first hint of his intention to retire came in a phrase in the October issue of the church magazine for 1959, in the letter which he addressed each month to the members and friends: he announced that this month marked the beginning of the twenty-fourth "and final year" of his ministry in the church. At once there were questions, and then in the letter for the following month he elaborated on his decision to go. The church, he declared, was in good health; and a great successful church in this new age had great opportunities. But the opportunities meant demands on the minister. In fact, he said, the demands were greater now than they had been in 1936 when he had first arrived in London: "and I am nearly a quarter of a century older. I find that after nearly twenty-four years' strain, some of them war years of shattering nervous demand, I now shrink from every new burden laid upon me and from every new demand which is made."

There was a good deal of opposition to his going, and even animosity among his most loyal friends; but he repeatedly harped on the three factors that determined the situation as he saw it—opportunity, demand, and declining power. He wanted to leave while he still had things going for him: " 'Go out with a bang and not with a whimper,' " he said, attributing the injunction to T. S. Eliot, who had in fact written roughly the opposite, Les being more reliable on the gospels or on Tennyson than on any modern poet!

The end of his formal ministry in London came twelve months after the first news of his impending retirement. By this time the church had chosen his successor, Leonard Griffith: the wheel had turned, upon which human affairs are ineluctably bound. In mid-October, on his sixty-seventh birthday, there was a farewell meeting, at which the church secretary praised him: he said the old things—that nothing he touched he did not adorn, and that no material gift could express the gratitude of the church. But material gifts there were, generous big ones: for Les, a cheque; for my mother a formidable pudding machine, with attachments and controls that enabled her to blend, fold, mix, stir or whip whatever raw materials she had a mind to exert technology upon. In receiving the gifts, Les opened his remarks with, "Behold, the idle rich!"

Two days later on the Sunday came the farewell services, the great queues forming early, the tense atmosphere, the music, the lessons, the prayers, and then the last sermon delivered in the accents that had cajoled, comforted, charmed, and brought new life to men and women through the years. After the sermon, they sang "Abide with me"; and then, last, the choir sang the "Nunc Dimittis", while Les stood in the apse with his back to the congregation, in front of the cross. After the service he was seen shaking hands with friends, members, and visitors and chatting with them and laughing in his familiar way.

12

They had bought a house in Bexhill-on-Sea which overlooked the Channel. It was large with many rooms and windows, and a sun loggia at the back, and a conservatory in the front, in which my mother had all sorts of curious little house-plants which she tended lovingly, peering over her spectacles to water them with a small, long-spouted can. Upstairs were four or five bedrooms, one of which was fitted out with shelves to become the study: Les had his huge oak roll-top desk in there and the old chair, in the soft contours of which so much had been written. On top of the shelves were the photographs of old friends, some now gone, Will Sangster, Billy Northridge, Bunnus, Dick Sheppard, Albert Schweitzer, Ern Appleyard, Frank Dorey, an old daguerreotype of my mother when she was slim and young, and a photo of Les with the Archbishop of Canterbury both roaring with laughter: "I was telling Billy Temple the story of the chapel keeper and the monkey," Les used to explain. The house had a large garden, beyond which was an Anglican church, St. Augustine's, standing in wide grounds, mostly lawns; there was a row of cottages for retired church people and in one corner a garden of remembrance, where, marked by a teak bench and some Peace roses, my mother's ashes now lie. In the garden there were fruit trees; and when my parents arrived there in October, these were heavy with apples and pears, a harvest to be gathered.

It will not for a moment be supposed that Les quite suddenly that autumn withdrew from the world of affairs and slept away the days in a deck chair in the garden with the *Christian World* over his head to keep off the sun. He slowed down, of course. The major burden, the responsibility of making and delivering two sermons a week, had been lifted, and that fact in itself gave him a wonderful sense of freedom. But the press clippings of the early sixties show him very much engaged with the religious life, and with national events that impinged on matters of morality or healing or death.

When he had been President of the Conference, he had spoken in favour of the unification of the churches. It had been his duty to nominate, along with the Archbishop of Canterbury, a committee to study Methodist-Anglican unity, and he had chosen Harold Roberts to lead the team. But for years before his presidency, he had spoken for unity, ridiculing the narrowness that fostered the separate dissenting groups, emphasising the wide disagreement within the Church of England itself, and calling not for agreement and compromise between things of belief and doctrine but for united action—against what all could agree to be evils—which would bring unity irrespective of mutual consent among theories.

Way back in 1932, along with Dorothy Round, the tennis champion, and Harold Larwood, the fast bowler, he had been one of the speakers at the Youth Rally during the Conference when Wesleyans, United Methodists and Primitive Methodists signed the Deed of Union. The occasion is mentioned in the autobiography of Lord Mackintosh, who presided over the rally and recalled Les, "crowned . . . with a luxurious crop of black hair". Now in 1963, short on hair but not on enthusiasm for this cause, he entered again into the lists in support of unity. He signed the manifesto of a group of Anglicans and Methodists who called themselves "Toward Anglo-Methodist Unity". Then, in "A Last Plea for Anglican Methodist Union", he put forward some of the arguments he had used in the past: part of the article ran thus:

> John Wesley, Pope John, General Booth, Albert Schweitzer, John Knox, St. Teresa, Gladys Aylward, Billy Graham, Harold Roberts and Kingsley Barrett all deserve the name "Christian". Yet no creed, unless it were almost a single sentence, no liturgy or form of worship, and no system of theology could be drawn up which would satisfy them all. Are the matters then, about which we disagree really of sufficient importance to give the on-looking world the impression that we are an army of factions fighting one another, instead of an army united against the only enemy, EVIL? We all love Christ and seek to serve Him.

Wesley had said, "Though we cannot think alike may we not love alike? May we not be of one heart though we are not of one opinion?" It was characteristic of Les to see things in this clear fundamental light, to ignore what he considered to be trivial complications of the simple and profound opposition between Christianity and evil. But people *do* cling to the putatively minor matters that have shaped their routine behaviour—these matters that obscure Les's grand division between black and white but which form the basis upon which faiths are built. Nothing came of all the efforts towards union, as everybody knows; in 1972 all hope of union with the Anglicans finally collapsed.

Shortly after his arrival in Bexhill, Les heard from the attorneys representing Penguin Books, Ltd. in the suit brought against it by the Crown on account of the publication of *Lady Chatterley's Lover*. Would he, they requested, appear on the stand as expert witness in support of their client and in favour of the novel, a copy of which they were sending under separate cover. He read it; then, he told me, he carried it out to the garden bonfire and consigned it. I can picture him: the solemn tread over the lawn, past the pear tree to the corner near the vegetable marrows, holding the book gingerly by the corner as it if were the carcass of a dead animal, watched quizzically by two curious starlings and a robin, dropping the book into the hot ashes and dusting off his hands in the small ritual purgation, in part in anger and in part so that he could say as he said to me, "I burned it! Out there on the bonfire!" He told the attorneys that he would prefer, if possible, to give expert witness for the prosecution.

A lot of people bought the book, however, and didn't burn it. They lined up in the street to buy it, old men and children among them. A million copies sold. It was incredible, Les wrote, that the purchasers were engaged with its literary merit; he regarded it as one more example of sexual stimulation such as publishers and show business had come increasingly to purvey to a vulnerable public, unaware of danger. The subject exercised him: every opportunity he had about that time he spoke up against it, frequently in vivid and figurative language.

Shortly after the Penguin Books trial, he wrote *The Times* a letter headed "A Nation in Danger", in which he demonstrated the moral laxity in adolescents by citing statistics about V.D. and illegitimate births. *Lady Chatterley* came up again; and with it a report that in a handbag check at a girls' school, 80 per cent of the sixth-formers were found to be equipped with contraceptives. This last piece of evidence gave rise to a good deal of scepticism and indignation. "Britain's kids are fine", Les was instructed in a headline in the popular press, over an article which documented its general assertion by praising their dress and blaming their imperfections on the H-Bomb. From *The Times* he learned he had been "rash". The story about the sixth-form girls proved in the end to be a bit of folklore which had surfaced here and there elsewhere, with percentages ranging up to 90. Les said he couldn't help it: that's what he had been told; and anyway the general situation was rank enough. A couple of years later, addressing the annual meeting of the Public Morality Council, he was again calling down imprecations on *Lady Chatterley* and other related examples of sexual liberalism. Undeterred by his previous experience with percentages, he now related that "eighteen girls went up to a certain university . . . Before Christmas thirteen had lost their virginity"; and again he cited the alarming V.D. figures. The nation, he felt, was being undermined; its salvation lay in the family, in family life, in education, and in censorship.

Early in his retirement, preaching in Manchester, Les had urged the scrapping of "old forms, phrases, and clichés": it was of dubious wisdom, he said, to use words in the service and then explain, "we don't really mean what the words appear to mean, but something quite different". It was his old hobby-horse. In 1962 he wrote to *The Times* again: this time he was out after a more significant document than *Lady Chatterley*—the Old Testament. "Recently," he wrote, speaking of the lesson for the day in a Church of England service, "we had read to us the story of the massacre by Joshua not only of his male enemies but their wives and children . . . Again and again," he goes on, "one would like to rise in Church after the Old Testament Lesson and say, 'My dear friends, do not pay any heed to the irrelevant nonsense which has just been read to you. It has no bearing whatever on the Christian religion. Nothing is true about God if it is out of harmony with the spirit of Jesus Christ.' " The sentiments, coming from Les, were not, of course, new: plain speaking he had demanded all his career and he himself had practised it; and the words and the manifest spirit of Christ had consistently been the illumination he had used when confronted by the recorded bad deeds of the Jews, the coils of theology, the perplexities of sin, pain, and death.

These concerns and others of their colour were now, however, gathering in his mind and, indeed in his notes. And they were to be issued in two books, *The Christian Agnostic*, in 1965, the last great book Les wrote, and *The Busy Man's Old Testament*, in 1971. The former enjoyed remarkable sales as a hardback, and when after two years it went into paperback it sold 5,000 the first week. In 1968 Les was the principal speaker at a luncheon sponsored by the *Yorkshire Post* to celebrate the centenary of the publishers, Hodder and Stoughton; by that time his own sales with Hodder had amounted to 500,000 copies. *The Busy Man's Old Testament* selects parts of the Old Testament that are relevant to men today and excises, a little cavalierly perhaps, some of the barbarous annals that Les found irrelevant or downright anti-Christian. It complements *The Christian Agnostic* but is a work of altogether smaller dimensions and need not detain us.

The Christian Agnostic begins, "I am an angry old man, and I feel I must get the fire out of my bones, as John Wesley would say, before I die." He is angry at the overlay—the creeds, ceremonies, doctrines, and mutual intolerances—which obscures the essentially clear and straightforward message of Christ: "Follow me." He proves also to be annoyed at the upstaging in the church of the personality of Christ by that of the "neurotic" St. Paul. And he is writing for the "Christian Agnostic" who, attracted by the spirit of Christ and its challenges, cannot commit himself to certain church observances and dogmas. One believes not that which is imposed as dogma, he says, but that which the mind leaps out to accept, which alone has *authority*. He describes one of a number of experiences that had convinced him of the existence of God: he was on a train passing through Vauxhall Bridge station on a Saturday evening on his way from Richmond College to fill in for a minister taken ill. The compartment was full . . . But the passage should be quoted in his own words:

> For a few seconds only, I suppose, the whole compartment was filled with light. This is the only way I know in which to describe the moment, for there was nothing to *see* at all. I felt caught up into some tremendous sense of being within a loving, triumphant and shining purpose. I never felt more humble. I never felt more exalted. A most curious, but overwhelming sense possessed me and filled me with ecstasy. I felt that all was well for all mankind . . . All men were shining and glorious beings who in the end would enter incredible joy. Beauty, music, joy, love immeasurable and a glory unspeakable, all this they would inherit. Of this they were heirs . . . All this happened over fifty years ago but even now I can see myself in the corner of that dingy, third-class compartment with the feeble lights of inverted gas mantles overhead and the Vauxhall platforms outside with milk cans standing there. In a few moments the glory had departed—all but one curious, lingering feeling. I *loved* everybody in that compartment.

In his chapter on Christ, Les deals specifically with things that are particularly enigmatic to the layman—the Virgin Birth, the miracles, the significance of His death, and the manner of the resurrection. This part of the book especially attracted the critics, who responded to it, according to their predispositions, with praise, amazement, concern, or distaste. The discussion of the Virgin Birth particularly engaged them: the matter was, Les said, to be held *sub judice*; but though it was of interest it was of no importance at all. He proceeds all the same to rehearse the circumstance of Joseph, referred to as father in the New Testament, and yet represented there also as shocked at Mary's pregnancy. He cites a theory he had read that when after the annunciation Mary entered the house of Zacharias, the priest of the temple, for three months, she did so in order to become pregnant by him of the Messiah in a "sacred marriage", a practice which, though disappearing, was still extant.

In writing about the Church, in fact the churches, he finds once again that the way towards union is not by unification of belief but by a "united front against every form of evil". At the same time, the beliefs expected of the Church of England and the prayers prescribed for baptisms, marriages, and funerals by the Book of Common Prayer are laid out under a cold rational light. What kind of God, he asks again, is it supposed that these prayers are addressing? The hymns and psalms in turn get their due criticism—Les, as we have seen, could be pretty impish in this kind of discussion. "Recently," he writes, "a dear old lady in the row in front of me at an Anglican church was lustily singing, 'Whoso privily slandereth his neighbour: him will I destroy' (Psalm 101.5), when though no doubt candidates for extinction were all around her, she appeared to have brought only her umbrella!" His chapter on the Bible emphasises its heterogeneity: it was myth, poetry, legend, history, prophecy, biography, allegory, parable, drama; and it was not all equally inspired. William Temple, Les claimed, was just as inspired as Paul; and T. S. Eliot more inspired than the author of the Song of Solomon. (I don't know how Eliot gained entry—Les used formerly to rank him with hymn writers who uttered obscurity and nonsense.) Writing of providence, he admitted there were occurrences that were pure accidents; but providence, if it be construed as a bond connecting us, in Paul Tillich's words, with "the fulfilling love", existed as a force in the universe. In discussing evil and sin he allows, without committing himself, the possibility of the existence of evil spirits: "It may be so." Sin, as Joseph Parker had said, was "a raised hand, a clenched fist and a blow in the face of God"; all the same, without softening the implications of the metaphor, Les wanted to "deliver a modern layman from false thinking about it based on Paul's less inspired and obsessional point of view . . ." In his chapter on death and survival, he gives some remarkable instances of "the blazing evidence of immortality" in Emerson's phrase, and then goes on to plead the case of euthanasia: for thousands of elderly people, with labouring breath, hardened arteries, permanent tiredness, and creaking joints and muscles, dying would make a difference "like that between riding a creaking pushbike up a steep hill on a rough road against a bitter wind, and then driving a new Rolls Royce on a perfect tarmac road in the Italian Riviera on a bright summer morning in June!" It's a telling contrast, and not thoroughly persuasive! Next there is a good deal of lore about judgment and hell that Les is unwilling to believe, holding before him always an image of Christ as an infallible guide to what and what not to believe about God. With the matter of reincarnation he is on thin ice and knows it. "I know that Jesus is nowhere reported as affirming reincarnation, but, though an argument from silence is precarious, he surely would have denied words which he believed to be false, as he did on another occasion, when life after death was under discussion." His argument in favour of reincarnation considers the injustice of life without the possibility of a second time round, but it dwells most on a number of rather sensational examples of infant prodigies and instances of *déjà vu*: Gianella de Marco, aged eight, conducted the London Philharmonic Orchestra; Danielle Salamon played Mozart at the age of four; Sir William Hamilton learned Hebrew at three. Cases of *déjà vu* are cited, among them this, coming to him in a letter from a mother writing of her son David:

> At the age of seven he went to Rome with me to visit my grandmother—who stayed in Italy a great deal. On this occasion we met a noted archaeologist who took us to see a newly excavated Roman villa on the outskirts of Naples. David became extremely excited, and ran about till he found a Roman bath. The edge of the bath was made of polished blue tiles about five inches square. These tiles were engraved with the signs of the Zodiac. David knelt down, and called out: "Here's our bath and our tiles—mine had a bull on it and the fish was Marcus'!" As he said the name "Marcus" he burst into floods of tears and called out: "Take me away, Mummy, take me away—it was all so terrible, I can't bear it."

On another occasion, visiting some caves in Guernsey, David said there was a further cave in which a prisoner, whom he named, had been walled up. The authorities denied the existence of a further cave, but they were prevailed upon to excavate and, doing so, discovered a walled-up doorway and the skeleton of a man, whose name appeared in ancient records as the boy had claimed. At the age of fourteen, in the British Museum, he approached a sarcophagus and announced that there were three initials on the underside of it. Asked if he could reproduce the initials, he drew three Egyptian birds. "'That was my name,' he said, 'but you weren't there then. I was a kind of inspector; I had to mark the coffins if they were satisfactory.'" After some bureaucratic resistance, the sarcophagus was inverted and the aforesaid sign was discerned.

The Christian Agnostic closes with a credo and some heightened prose: ". . . in a tiny, imperfect way, I have known God. I am certain of God and that God is love." Much remains to be understood and the way is far. Sometimes, he says, he has ignored God and known bleak days.

> And then, warmed of heart by some contact with love, by some word of another, printed, written or spoken, or even by the glory of a summer morning, or the quiet music of the sea at night, I come back, and his grace comes flooding in, cleansing, refreshing, renewing, uplifting, like the sparkling tide in a dull, muddy backwater. Then the whole earth

fills with his glory, and once more I know that all is well. The mind recovers its resilience and gaiety, the soul finds its rest, and the will takes once more the upward road, and *knows* that the Great Companion is near, and that he will never weary of mankind, or regard a single soul as too unimportant, or too unworthy, to be his friend.

The book was a summary of his life's work: the beliefs that had nourished him and that he in turn had preached for fifty years. It met with the mixed reception that he had become used to: hostility and rejection from the fundamentalists and the inflexibly orthodox; a cool dissection and some pained criticism and dismay from theologians; admiration from doctors and lawyers; admiration and immense gratitude from the ordinary men and women of a thousand different walks of life for whom the book had been largely written.

The publication of the book coincided with the fiftieth anniversary of Les's entry into the Methodist ministry. In fifty years he had stirred up some controversy: since *After Death* and the charges of heresy that it had incurred he had not ceased to flutter the dovecotes in one or another precinct of Methodism. Now, however, he was widely known and something of a figurehead, and the things he said were a gauntlet thrown down in a national arena. Many wished to pick it up: there were the simple counter-assertions assumed to be self-evident: one reviewer, commenting on Les's interpretation of the Cross, wrote quite dogmatically, "Without shedding of blood there is no remission of sins"—So *there*, so to speak! There were those who deplored investigation, any investigation *per se*, and wanted to be left with the "awe, mystery and wonder of the faith". There were a number who objected (as their predecessors had objected to *After Death*) to Les's taste, which preferred Robert Browning to St. Paul, and Alice Meynell to the Athanasian Creed, and which was willing to accept the poets as superior authorities. There were the reviewers who played one-up-manship with their formidable fire power: "Can one seriously confront an educated 'Christian agnostic' of today and not refer to de Lubac, Teilhard du Chardin, Hans Urs von Balthasar, Karl Rabner, Hans Kung and the rest?" They forgot Pico della Mirandola.

Most of the adverse criticism of the book was directed against the subjectivity of many of its interpretations and against the specific interpretation, which was not subjective, of the texts about Christ's birth. Les's criterion for authority met with criticism: with the formula that what he perceived to be true alone had authority, he had made up his own system of belief that was supported by an eclectic choice of scriptural and other materials. He had, for instance, given credence to some passages and withheld it from others in a single gospel. By such a process, a reviewer pointed out, he might have found any political creed to be "true". He had, it was charged, tried to remove from doctrine those things that men found hard to believe: it was the strategy Ronald Knox had condemned—rewriting belief on the principle of "How much will Jones swallow?" "One feels," said another reviewer, "that the universal love he looks for is universal pleasantness." At the same time it was claimed that, while he was reducing the difficulties attendant upon orthodoxy, he was adding new ones.

His discussion about the Virgin Birth was one such case; but with this, critics found his alternatives not so much difficult as disrespectful. The idea of Zacharias as father of Christ that Les cites as a possibility, without endorsing it, drew forth a good deal of spirited denunciation. One reviewer writes, "It is highly speculative and . . . will distress a good many devout people." Among the latter was Ian Paisley, whose name appears as author of a small pamphlet entitled "Judas Iscariot, Elymas the Sorcerer and Methodist Leslie D. Weatherhead: Three of a Kind: An Exposure of 'the Damnable Heresies and Doctrines of Devils' Set forth by Dr. Weatherhead in his latest book 'The Christian Agnostic' " in which a number of passages from the book are quoted, commented on and countered by passages from the Bible. The idea of the fatherhood of Zacharias calls forth some eloquence: "Such blasphemous immorality could only be conceived in the satanic mind and propagated by an emissary of the lowest hell. Only a vile brat of the pit of perdition could dare to set down such an alternative." Les certainly never expected unqualified praise from all quarters, though I can remember once that he did register some surprise in Shaftesbury Avenue when a taxi-driver leaned out of his cab and called him an obscene name! I don't think the adverse criticism of his book bothered him unduly—certainly not the Reverend Paisley's Christian response, and not, for that matter, the review that closed, "You will like this old warrior, and you will agree with what he wants to say and do. We need this kind of bull in our neat little intellectual china shops. But it is still bull." He responded in one or two cases to reviews that misquoted and hence distorted one passage or another. And when the *Times Literary Supplement* reviewer pronounced that "If the preacher thoroughly presented Dr. Weatherhead's version of Christianity, he would undoubtedly lose most of his congregation", Les could not resist instructing him that the City Temple had been full for twenty-four years.

Much of the criticism of the book was introduced by tributes to the life-work of which it was the summation. All in all, even where his theology was under fire, there was a good deal of respect. He took most of the criticism in his stride: he could hardly be surprised that fundamentalists did not mutely sustain his liberal rewriting of the Word or that theologians should resist the lure of publicly debating some of the issues. The favourable reviews, in any case, far outweighed the others. It was not in fact for people committed, not for theologians, not, certainly, for those lords of orthodoxy who wielded the power of the Index that he had undertaken to retell the story. But, in fact, one of the most surprising tributes appeared in a Catholic monthly.

> I picked it up [wrote the reviewer] without any great excitement and began a rather desultory reading. Very soon, however, my attention was caught and I read it during every spare moment in a state I can only call "thrill". The book, as the title suggests, is anything but orthodox and Catholic, but every page in it seemed to me to breathe the love of Christ, the vital importance of religion, a sense of constructive freedom, and a genuine spiritual enthusiasm with which one very rarely meets.

He is writing, he begs his readers to note, for subscribers to this Catholic magazine; in general circulation the book would call, he says, for "the sternest blasts of the Index Expurgatorius". "Yet," he goes on, "to me at least, the love of Our Lord and the dedication to truth and common sense, as Dr. Weatherhead sees it, make an overwhelming appeal particularly at the present time, when the Catholic Church is in process of Aggiornamento and reform. Let me put to one side what we really cannot, by our Catholic faith, accept, and concentrate on what it is that makes the Methodist Dr. Weatherhead so much better seeming a Christian than I and perhaps you are."

The Catholic Church was perhaps the last place from which Les would have expected his work to be received with such warmth. The men and women he had in mind as readers were the ones represented, for example, by Mr. Ian G. Gregory, who responded to the criticisms he had read in the religious press with a short postcard saying, "We are astonished at the rot being written about your new book by people who obviously have a vested interest in traditional religion. On behalf of the salesmen, water engineers, accountants, teachers, centre lathe turners, window cleaners, housewives, schoolboys and girls and me, a journalist, thank you, thank you, thank you . . ." A lot of the salesmen and water engineers, however, and their wives, too, wrote on their own account, and hundreds of letters flowed in. "I am one of the persons for whom you wrote the book," they said; they said the book embodied many of the thoughts that they had been entertaining all their lives. "I am only a housewife," one letter said; "imagine my suprise when I read my very own thoughts in your book"; another, "These things I had known since I was young." On seeing their ideas reproduced here and confirmed, a rather remarkable number of people found alleviation of the feelings of guilt that they had nourished on account of the unconventionality of their Christianity. The book brought them peace of mind; it answered questions that had perplexed them; it was a catharsis of feeling that had been buried and unexpressed. One letter described how religion "suddenly all came alive"; another declared with manifest relief, "At long last after years of searching I can say I *am* a Christian."

In his first years of retirement Les took on many preaching and lecturing assignments. He engaged in controversies, as we have seen, over such questions as sexual morality and church unity. He made a lot of appearances at prizegivings and in Bexhill he again became chairman of the governors of a girls' boarding-school. He became a familiar figure in the little Sussex watering-place: he laid wreaths on war memorials on Armistice Day and preached in the Methodist church. And in the greengrocer's store and the stationer's they asked after his health and agreed that he was getting on a bit now. He accepted a few patients referred by the doctor or one of the local ministers and gave them psychological help. He took an interest in the establishment of a Methodist Home for the Aged in Bexhill; and when it was completed it was called "Richmond", partly in honour of Les who had been at Richmond College and now lived in a street called Richmond Grove; he opened it and donated a thousand pounds for a lift in memory of my mother. He took some pleasure in his garden, consulting with the gardener and doing some of the less skilled chores like weeding and tending the bonfire (which he did not regularly feed with literary items). He used to love to drive with my mother through the narrow lanes of the county, picking primroses and violets in their season. Or they would drive through Ashburnham Place, near Battle, with its superb beeches and cedars. Or towards the end of an afternoon they might walk a quarter of a mile and lean over a gate. Les might quote old Tennyson: "The trees laid their dark arms about the field"—he would be quite likely to do that. And then they would putter home through the dusk. In the evening they did the *Evening Standard* crossword, watched the telly, played Scrabble, and went early to bed. A good life: their marriage had never meant so much to them.

Les would make fairly frequent excursions to London, to preach or to go to the theatre. My mother, who had never mastered her fear and dislike of public transport, rarely accompanied him. But Norman and Hansie French would meet him at the station in their big blue car and install him in their hotel and tender him every conceivable comfort. He still loved the theatre: they took him to a Pinter play, and he let them know that he could have done better himself; but the regular old thing, he enjoyed. "I'm going to see a play tonight with Norman," he wrote to me. "It's by a chap from my old school called C. P. Snow." Plays by chaps like Snow engaged him tremendously; and as they drove home, he would rehearse the tenser moments: "I say, wasn't it terrific when that fellow climbed in the window?" (or whatever it may have been). Or "Why would he want to do all that for that mousey-looking girl?"

In the spring of 1968 I drove with Les to St. Margaret's Bay, where the Dial House had been sold long ago. It was to be a three-day holiday, staying at the Bay Hotel, visiting some of the old haunts and indulging some of the old memories. The first day we spent on the marshes watching the herons flopping about in the trees. We walked a bit, had a ham sandwich and some cider in a pub, and drove back to the hotel for dinner. After dinner in the dusk we took the car up to the lighthouse and watched the shipping making for Dover harbour. Then, pretty early, back to bed. A reasonably easy day; Les was seventy-five. In the night, however, he had a heart attack. I got the doctor and we moved him into hospital. The attack was not massive: the doctors in their knowing way called it an adventure, arguing over a little kink in the electrocardiogram, poring over it as if it were a hieroglyph in the papyrus manuscript of the Pentateuch. Les was cheerful when I visited him: he apologised that he'd given the roses I had sent him to a pretty nurse called Angela, and he did an uproarious imitation of the orderly in the hospital who hadn't any teeth. But the incident was a milestone: now he knew he had a "heart condition"; he must take care; he had arrived at a new and lower plateau in the downhill journey. After this he no longer dared to preach, fearing he might have a seizure in the pulpit, and he shortly ceased to take any part in church services.

During the next two years my mother suffered one or two minor strokes. Her speech became slow and her handwriting small. Knowing there was no cure, she laughed at these infirmities, apologising for her writing and joining in the general amusement when we told her that it had always been henscratch anyway. She was left at length without the ability to talk; she longed to die. She had in the end to be attended day and night, and she made signs pleading not to be sent to a nursing-home. She had always been a home person, hating even good hotels, loathing hospitals; Les determined to spend without stinting to give her what comfort could be had as long as his savings lasted. He hired nurses to look after her at all times. They were bad days for both of them.

"When did you last wash your hands?" I heard him demand of one of the nurses who was giving my mother a pill. I warned him, "They'll quit on you if you don't look out!" "Not they," he replied in his old stage whisper. "Not at the rate I'm paying them!" The house was upset: there were always things being carried upstairs; there were afternoons with voices on the landing; and there were obscene-looking apparatuses and the smell of disinfectant in the bathroom.

Friends had given my mother a steel board with magnetic letters with which she could spell out words and sentences. She moved the letters slowly and deliberately with her large hands. When the drift of a sentence became clear we would speak the end of it for her. "All right," Les would say, "we've got it. You don't need to go on." But she would insist on finishing the sentence, laboriously moving her fingers, running out of letters and borrowing them from earlier words; her will remained unimpaired. In this manner she asked me about a gift she and Les had sent us in the States which I warmly acknowledged. "Les chose it and bought it by himself," she then spelled out. This may well have been the last complete statement she made to me; and if so it was characteristic that it should have been concerned to give credit to her husband with the kind of generosity and self-abnegation of which she had made a practice throughout her life.

The family gathered at the funeral—Margaret from Welwyn Garden City, Dick from Ohio; I came from London. Les rallied a little with his sons and daughter around him, and the release of my mother from her suffering took a burden likewise from him. But he was slow to recover the strength which had been depleted by the last sad months of her life; his health was not good, not even for one in his late seventies. He suffered pain from diverticulitis, from the skin rash, and from rheumatism. Minor concerns gave him untoward anxiety: he worried, I remember, about the new coinage that was to come in in the early 1970s and about the introduction of North Sea gas which he feared would blow them all up. He was lonely. Even with the extraordinary and selfless devotion of Gertie, the housekeeper, who continued to look after him, life was unavoidably empty. He didn't want to write any more books; he dared not preach; he no longer felt well enough to go to London. He had had his life, he felt, and now he was impatient to be translated into the life beyond death in which his belief had never faltered. "I long to slip away," he wrote in these months. "There are so many people waiting for me on the other side."

His routine has now become small. Morning tea is brought up at 7.15; he breakfasts at a quarter to nine. After breakfast, he carries a little tin basin of scraps out to the bird-table in the garden, watched greedily by the gulls from the roof of the church and the starlings on his own television aerial. After feeding the birds he repairs to the study for a while to write a few letters, read the *Daily Telegraph* and the religious papers. He prays a great deal. At eleven there is his morning Bovril. At one he eats lunch and hears the B.B.C. news. After lunch he returns to the study, eats two squares of Terry's Bitter Chocolate, and sleeps. There is a cup of tea at three-fifteen and then he takes his stick and walks round the block, or sometimes walks to the sea-front and sits on the bench near the old mine in which one can put pennies for the lifeboat men. There he is out of the wind and can watch the sea. When he walks he stands erect: a whiteheaded prophet left over from the old days, looking down from every inch of him upon the incredible follies of a world in which he feels he no longer has any part. At five-thirty there is tea. And then after that, he and Gertie will play Scrabble until nine when the T.V. news comes on. At nine-thirty he goes up to bed; and, before sleeping, for an hour or so he will re-read the old Sherlock Holmes stories that had engaged him as a boy.

A quiet life. Neighbours salute him; old friends occasionally telephone and occasionally call at the house. Letters from admirers do not cease to trickle in, and he answers them in the same old green ink. His mind remains clear. In the spring bronze wallflowers bloom near the front porch, and in the autumn the dahlias against the hedge are a splendid show; the gardener keeps the lawn like velvet. Gertie brings the meals on time, perfectly prepared. And these things are a comfort, but only mitigations of a lingering exile for him to whom death is wide and loving arms.